C000272433

DOGS
MISBEHAVING

BY THE SAME AUTHORS
Bach Flower Remedies for Horses and Riders (Kenilworth Press)

DOGS MISBEHAVING

Solving problem behaviour with Bach Flower and other remedies

Martin J. Scott and Gael Mariani

FOREWORD BY JULIE SELLORS

KENILWORTH PRESS

First published in Great Britain 2001 by
Kenilworth Press
Addington
Buckingham
MK18 2JR

British Library Cataloguing in Publication Data
A CIP record for this book is available from the British Library

ISBN 1-872119-40-9

Illustrations by Pam Tanzey

Typesetting and layout by Kenilworth Press
Printed and bound in Great Britain by South Western Printers, Caerphilly

Contents

Acknowledgments

A big thanks to all the people and dogs who have helped with the development of this book:

to the flower essence producers, whose wonderful remedies are changing the lives of so many dogs and their owners; we would like personally to thank Steve Johnson of the Alaskan Flower Essence Project, Patricia Kaminski and Richard Katz of the Flower Essence Society in Nevada City, California, Ian White of the Australian Bush Flower Essences, Julian Barnard at Healing Herbs Ltd, and the good folks at Crystal Herbs. And last but not least, we pay tribute to the memory of Dr Edward Bach.

to the late John Fisher, whose work has been so inspirational;

to Julie Sellors, for her kind support and input;

to Pam Tanzey, for her wonderful drawings;

to Patty Smith-Verspoor, Rudi Verspoor, Mary Rothschild, Thomas Mary D'illon at FES, and Don Dennis and the staff at the Living Tree;

to Lesley Gowers and all the team at Kenilworth Press who are such a pleasure to work with;

to Liz and Gordon at Great Dane Care in Carmarthen, Wales;

to Joanne Vyse, for her enthusiasm in spreading the word;

to Sebastiano Puglisi DVM, MRCVS, our great holistic vet;

to all the many dogs and owners we have worked with along the way, who have taught us so much about our craft;

and lastly, to Gaia and the spirit of nature that lives on in the flower remedies and makes it all possible.

Foreword
by Julie Sellors

In recent years there has been an interest in developing kinder, more effective dog training methods based on an understanding of our dogs' behaviour. Through this understanding a deeper appreciation of our dogs' emotional state, and the part it plays in our relationship with them, has been recognised.

These developments naturally lead us to seek ways of not only modifying our dogs' behaviour, but also of assisting when their emotional state is out of balance. Through this book, Gael and Martin share their knowledge and experience in doing just that. The detailed information means even people new to flower remedies can select suitable remedies to help their own dogs, and the benefits of flower remedies are illustrated in case histories.

This book is informative, interesting and thought provoking and I am sure that you, like me, will enjoy reading it. Once read, it becomes a useful reference tool to help you when your dog needs that additional emotional support or adjustment which the remedies can bring.

Julie Sellors is a professional animal behaviourist and a member of the APBC (Association of Pet Behaviour Counsellors). She is a behaviour consultant for the Blue Cross, Burford, Oxfordshire, and runs her own private practice. She is a tutor on the prestigious John Rogerson Canine Partnership Course, along with John Rogerson and Gwen Bailey. Julie has contributed to many books on canine behaviour, and is a specialist in Contact Learning.

*This book is dedicated to
Yoda, Mungo, Pagan and Ziggie,
the best teachers we have had.*

1 What are Flower Remedies?

Few things in this world are as simple as flower remedies, yet few things are so astonishing. They are a sensational example of the healing powers of nature, a medical hypertechnology that unites ancient wisdom with the very latest advances in modern biological scientific research.

Perhaps, like millions of people around the world, you are already acquainted with flower remedies, in which case you may have a pretty good idea of their uses. If not, they are really very straightforward. Flower remedies are a type of natural healing therapy. Unlike some therapies, such as homeopathy, they aren't used to cure physical illness directly, but instead focus on helping to heal **negative emotions**.

We all know a negative emotion when we see (or feel) one: things like grief; wounded pride; shock or trauma; anxiety; terror; rage; bitterness; guilt; rejection; feelings of futility and emptiness; loss of determination, drive or confidence. It could be a long list, but you get the idea. All of these negative states of mind, which may come on suddenly (acute), or may develop gradually and become entrenched in our mind and drag on for months or years (chronic), or may also be part of our actual personality, can be serious blockages to happiness, enjoyment of life, personal progress and even physical health.

What is so great about flower remedies?

• Flower remedies offer us a safe, natural means of overcoming these problems, freeing up our potential for happiness. They have the ability to transform negative states of mind and mood into positive ones, bringing about an emotional re-balancing and allowing our own better virtues and courage, previously hidden, to emerge. So if I take a flower remedy for, say, fear, the courage I begin to display as a result is my own courage and not just the artificial effect of a chemical introduced to my system.

- Radically opposite to conventional medicines such as antidepressants, they do not attempt to stamp out or numb the symptoms of emotional pain (chemical **suppression** is actually considered by many homoeopaths and other health professionals to be at best crude, at worst downright dangerous). Instead, flower remedies gently get to the root causes of such pain and iron out the imbalances, working, so to speak, from within.

- They are not drugs, and not chemical in nature.

- They are pure and non-toxic. There have been **no** known cases where flower remedies proved less than a hundred per cent safe.

- They are extremely easy to use, and if the wrong one is chosen by mistake it simply does not act. It is impossible to cause problems by giving the wrong remedy, or by giving too high a dose.

- All of the above has been systematically tested and verified over a period of almost seventy years. This is not some new gimmick!

You can buy flower remedies everywhere: health stores, mainstream pharmacies, specialist producers and dealers on the Internet. They are quite inexpensive to purchase, last a long time, and keep more or less indefinitely. They come in liquid form in little amber bottles, from which only a few drops are taken orally each day, either 'neat' or added to water or food (we shall go into all the practicalities later). They are rapidly becoming one of the big industries of the world-wide boom in complementary therapy, and people from all walks of life, in all countries, are beginning to discover them. We may not yet realise it, but we are entering into a new era of health awareness, and flower remedies are going to be a big part of the future.

The origin of flower remedies

It all started with an English doctor, Edward Bach (1886-1936). Bach was a bacteriologist turned homoeopath, who became convinced through his clinical practice that the state of the emotions was a crucially overlooked factor in the treatment of all disease. He was a tirelessly inquisitive and systematic scientific pioneer, and also a highly spiritual man with a deep reverence for nature. He quit conventional practice and his offices in London's fashionable Harley Street in 1930, and spent the last years of his life seeking out the 'healing herbs' that would between them help treat the entire spectrum of negative emotions. After much experimentation and research, and even more time spent walking around the wild parts of England and Wales in search of curative wild flowers, he finalised the set of thirty-eight remedies plus one combination remedy that have come to be known across the world and marketed as the Bach Flower Remedies.

The process Bach invented for preparing these natural elixirs is very simple. A few flowers of a given type are picked when at the peak of their bloom cycle, placed in a bowl of spring water and left to sit in the sun for a few hours. The sunlight allows an energising or **potentising** process to take place, whereby the healing properties of the plant pass into the water, where they are stored or 'recorded' at a sub-molecular, energetic level. Then brandy is added as a preservative, not for the plant energy but for the water, and the flower remedy is ready for bottling. Its energy will live on long after the material substance of the plant has died away and decomposed. In fact some of the original flower remedies made up by Dr Bach are still preserved.

Probably the best known Bach remedy is the famous Rescue Remedy, a combination of five flower remedies, which is routinely used as a general emotional cure-all and shock reliever by millions of people. Many people swear by the Bach remedies and Rescue Remedy in particular, yet they may not be aware that the little bottle they carry in their pocket or bag is a mind-blowing technological wonder – the 'vibrational' or subtle-energy imprint of a mixture of five wild flowers whose workings on the psyche are so high-flown that it is only recently that modern science, through a marriage of radical physics and enlightened biology, has been able to begin to explain them.*

Development of the international scene

For many years after Bach's time, the field of flower remedy research remained in stasis. The promotion of flower remedies remained largely in the hands of the Bach Centre in Oxfordshire, the team of people who dedicated themselves to preserving Bach's legacy. The Bach Centre has done much good work since 1936. A great number of case studies were carried out and documented by these devotees, the production of the remedies was continued and eventually passed on to a bigger company, Nelsons, as demand grew, and a number of books were written by the Bach Centre team which form a good basic introduction to this family of flower remedies.

However, there was never any attempt to develop Bach's work any further, or to work on creating any further remedies in the same vein; the Bach 'system' was strictly considered a closed loop, a fully organic, self-contained and exclusive therapy in its own right. The attitude, carved in stone, seemed to be that Bach had single-handedly discovered every single healing flower in the world, and Mother Nature's global plant kingdom had nothing further to offer in the flower remedy department.

It was inevitable and necessary that, sooner or later, further development of flower remedy therapy would take place. Slowly, over the years, many professionals in various parts of the world who had been using the Bach remedies and seen their benefits first-hand, began to question why other plants in other countries shouldn't have similar potential to those discovered by Bach in Britain. It seemed obvious that the therapy could and should be much more international in scope, especially considering the extreme wealth and lushness of flora in warmer climates and more unspoilt ecosystems.

---

* Dr Keith Scott-Mumby's book, *Virtual Medicine,* gives a clear account of how modern science is rapidly moving towards a full validation of subtle-energy medicine, much to the discomfort of the medical profession.

And it was in the sun-drenched meadows of California and the sprawling, untouched wilds of Australia that two of the more prominent modern families of flower remedies were born in the 1970s and 1980s. They were developed and researched carefully, systematically, over a period of years by their originators, Patricia Kaminski and Richard Katz in the USA and Ian and Kristin White in Australia. It must be understood that while these new products are created along exactly the same lines and work in the same way as the Bach flower remedies, they are not claimed to be 'new Bach remedies'. With these new families of remedies a new term began to come into usage: flower **essences**, referring to the nature of the remedy as the subtle-energy essence of the plant. And so we have the Californian Flower Essence Services (FES) collection, and the Australian Bush Flower Essences, respectively. The term 'essences' is now very widely used and is quite interchangeable with the rather more old-fashioned term 'remedies'. Please don't be confused by the evolving terminology.

The bountiful, beautiful flora of Alaska have brought us another range of flower remedies, or essences, developed by Steve Johnson, founder and director of the Alaskan Flower Essence Project. There are additional producers of flower remedies in other parts of the USA, in Canada, Brazil, the UK, Ireland, France, the Netherlands, Australia, Asia... everywhere! (See the list of suppliers at the back of the book.) The wealth of flower remedies at our disposal is now huge, ranging into many hundreds of different remedies, each with its own specific use for treating some particular shade of negative emotion for both people and animals.

We are quite certain that Edward Bach, if he were still with us, would strongly approve of the growth in flower essence research and therapy at the end of the twentieth century. It is partly thanks to the arrival of all these new remedies over the last quarter of the century that a consumer boom has been sparked in recent years and flower remedies are rapidly gaining interest and popularity. The Australian Bush Flower Essences, to name but one source, sell in huge quantities in the UK and the USA as well as their native country.

But apart from being a business, flower essences are a *science*.

And like all science, they have to be consistently proven and researched. For that reason, professional practitioners and students of this type of therapy are slow to accept new flower remedies into their repertoire. Only those remedies that have been around for enough years to provide detailed, consistent, controlled evidence of their efficacy have been fully embraced, which is as it should be. For that reason, as well as the Bach remedies themselves, this book deals only with the modern flower remedies that we ourselves have seen working in our own experience.

The remedy families we know, use and write about in this book are the original Bach remedies, the Australian Bush Flower Essences, the Californian FES products and some of the Alaskan. That way we are assured that you, the reader, are getting reliable information 'straight from the horse's mouth', so to speak. This is in no way to doubt or denigrate any of the other families of flower remedies, many of which have been around a good long time and been widely used. But we feel it would be wrong to write about remedies we have not yet used ourselves. As our own research on animal therapy progresses, we will surely come around to it – if one of our many colleagues in flower essence therapy does not get there first!

2 Flower Remedies and Dog Behaviour

Many people are already switched on to using flower remedies in their daily lives, to help them cope with problems like stress, sadness, grief, impatience, anger, frustration, guilt, disappointment, jealousy... the list is endless, ranging from every-day mild problems to more deep-seated chronic emotional imbalances that cause great suffering.

But what about animals? Some people are surprised at the idea that a type of therapy which treats human emotional problems can also be used effectively on our canine companions. Yet, really, it should come as no surprise at all. Even non-dog owners are perfectly aware that dogs are capable of different emotional states. At the most basic level, a dog's frame of mind is what decides whether he sits quietly at his garden gate as you walk past on your way to work in the morning, or whether he launches himself at you in an attempt to bite your leg!

Those of us who live with dogs, share our homes with them and look after them, are able to perceive them as very complex emotional beings that are capable of a wide range of moods and feelings. They have the ability to make it quite clear to us when they are happy – and few things are as warming to witness as a bright-eyed, contented dog rolling ecstatically on the grass. They can show great affection to each other and to us. They can show us when they are miserable, by withdrawing from company and preferring to be alone. They can sulk when feeling sorry for themselves. They can manipulate us by turning on the charm when necessary. They can be devious and calculating, waiting for our backs to be turned to steal something or bring out some forbidden toy secreted away in a private place. They can be demanding, prone to temper outbursts. They can get jealous and resentful. They can anticipate pain and punishment, suffer extremes of anxiety and fear, and be traumatised for life by abuse, cruelty and neglect. They can also suffer great pain and sadness after the death of an owner or companion.

In short, dogs have access to a range of mental and emotional states that is comparable to our own. In fact the emotional capacity of a dog is very reminiscent of that of a human child. And above all else, just like humans, as any experienced 'dog person' will tell you, each dog is marked by his or her own individual character, which to a large extent dominates and dictates its behaviour, mood patterns, the way it fits in with both canine and human society, and its responses to such stimuli as training.

But just because dogs share some of our human emotional characteristics, does that mean they're just the same as we are?

Trainers and healers

There are two distinct schools of thought when it comes to dog behaviour.

One, probably the most prevalent, is the school of dog **training**. Dog training is, in principle if not always in practice, the art of using gentle, non-confrontational methods of positive and negative reinforcement to shape and mould a dog's behaviour so that it is happiest when in tune with what we require from it, i.e. that it fits in with the human society to which it has been introduced. Purists of the dog training world would not deny that dogs are capable of emotional shifts and states, but they would take the view that these emotional characteristics manifest themselves as **behaviour**, and behaviour can be altered and modified by training techniques – therefore training is the be-all and end-all and there is no need for any further approach.

The late John Fisher was one of the world's leading authorities on dog behaviour, psychology and training, and a highly respected man indeed. A founder member of the Association of Pet Behaviour Consultants and the author of several books including *Think Dog!* and *Why Does My Dog…?*, Fisher was a full-time professional canine behaviourist with clinics in London and Surrey. He worked closely with many official bodies on dog behavioural problems, and was a top canine psychology lecturer and educator until his tragic and untimely death in 1997. We can assume that, when it came to dogs and dog behaviour, the way dogs think and view the world, John Fisher knew what he was talking about.

Yet Fisher was quite aware that, for all its usefulness and sophistication, the modern system of dog training was clearly not the whole answer when tackling problems of canine behaviour. There were times when conventional dog training methods were simply not enough to deal with the situation on their own. For instance, what is a trainer to do when confronted with a dog that is so aggressive that it cannot be approached? Or a dog that is so afraid of people, perhaps after having been cruelly treated, that the mere sight of someone approaching makes it urinate in terror? How is the dog to overcome its trauma and realise that we are trying to help and that we are not just another tormentor? How are we to heal the frightening memory of past beatings in a dog that cringes pitifully when we innocently raise an arm to scratch our head or take something off a shelf? Or what about dogs that simply do not respond to the lessons of training and will not let go of negative tendencies that may be ingrained into their personality from birth? (Moreover, what is a trainer to do with an owner who professes to be afraid of their dog, perhaps too afraid to do any training with it?)

Fisher came to realise that other methods could and should be introduced to the system in order to produce a more balanced and complete approach. He advocated the use of the Bach remedies, and writes in *Think Dog* that he found their effects on problem animals 'mind-blowing'. He was able to solve cases of problem behaviour quickly and effectively, using a combination of his skills as a trainer and correct use of flower remedies.

Fisher struck a rich vein here. The pure training approach, focusing on **behaviour** rather than **emotion** – effect rather than cause – runs the risk, in certain cases, of leaving untouched and unhealed many of the underlying emotional disturbances, character weaknesses, mental wounds and traumas that dogs can experience. A person who had been traumatised or abused, and reflected that in their behaviour, would be offered therapy rather than 'behaviour modification', so why deny that option to a dog when flower remedies offer us the means to do so?

This leads us on to the other school of thought, one that is often adhered to by practitioners of natural medicine. Books have been written on using flower remedies for animals which fully take into account the capacity for dogs and other animals to experience a wide range of emotional problems, and give guidelines on how remedies may be used to heal these **in exactly the same way as would be done for people.** We call this the emotional healing approach. The predominant attitude here is pretty much the converse of that of the dog training school. Where the purist trainers say that all emotions show up as behaviour, and thus behaviour modification is the key to the problem, the emotional healers take the view that all behaviour is born of emotions, and thus healing the emotion with a remedy such as a flower essence is the only way to get to the problem.

This emotional healing approach holds true much of the time. As canine behaviour therapists, we get a lot of calls from people with problems involving animals that have become afraid after suffering a shock to the system, and we have found that these respond far better and more quickly to flower remedies than to regular training methods. In fact many of these dogs have been seen by qualified trainers before they come to us.

Many people take in rescue dogs, such as ex-puppy-farm dogs, dogs that have been abandoned or come from unsuitable homes, and find that they are emotionally scarred. It is often seemingly impossible to re-integrate them into normal society, and the traumatic disturbances in their unconscious time-line are so deep that they can never be happy again, often never trusted again. It might take many months of very expert and sophisticated (and expensive!) training to iron out problems like that, but flower remedies can often manage it within startlingly short periods of time – sometimes just a few days, or even less. Amazing stuff! (See the case histories later in the book.)

But the emotional healing approach, viewing all behaviour as treatable purely via the emotions, also has its limitations. One of the more common reasons for our phone to ring is the 'naughty dog'. 'Naughty dog' covers anything from dogs that steal or behave aggressively to their owners, to dogs that bark continually while their owners are on the phone or watching TV. Now, here, we are not quite so quick to reach straight for the flower remedies. When we tell the owners that the problem has been aggravated, maybe even created, by **their** behaviour rather than the dog's, we get a lot of surprised reactions. They no doubt expected to be given a magic remedy that cures NDS, 'Naughty Dog Syndrome'; instead they have to be filled in on the dos and don'ts of basic dog psychology and training. We have to explain to them that while the flower remedies will work wonders on problems that are purely emotional, such as after-effects of abuse or fears stemming from particular traumatic incidents such as fights with other dogs, being hit by a car, etc., they cannot be expected to perform miracles in cases where a dog is only reacting to its environment according to its natural instincts and values. It is, after all, a dog and not a human. To understand what we mean here, let's take a brief look at the dog's view of the world.

The pack

The dog is an animal that requires membership of a social group, an interdependent team, called the **pack**. We may have bred dogs into all shapes and sizes, but at the heart of each one, whether a Shi-Tzu or a Great Dane, is the instinct of a wild hunting wolf that relies on the strength of the pack for its survival. Living within human society, dogs are quite happy to regard us as honorary pack members, and the way they relate to their owners and everyone within the household is dictated by their pack instinct. So, just when we thought we'd brought an animal into **our** world, we see that in fact the dog is still very much in the same world he has always occupied and it is we who have been dragged into a new world. First you were a **family**, but when you got a dog you became a **pack**!

Now, the thing about packs is that they need a pack leader. If the dog doesn't feel that the pack is strongly enough led, to him

'The pack.'

that is a sign that it is vulnerable, and his instinct is to rise up and take charge... strictly for your own good, you understand. There is no malice in it. The pack leader, or Alpha, accepts much responsibility for the safety of the pack, and in return gets all the privileges: he always leads the way, always eats first, is never disturbed and always has his pick of where to sit, sleep, and so on.

The dog can never understand our strange human values; it is we who must try to understand his. The onus is on us if we want to create a good relationship between man and dog. If we fail to take on board the dog's own values and the importance of pack politics, we risk creating problems. Many dogs consider themselves pack leaders within a household, without you being aware of it. Consider the following scenarios:

• When you let your dog sleep on your bed, you are allowing him to contest your position as leader. No pack leader would allow its privileged space to be occupied by an underling: therefore, by simple logic, you obviously are not a very strong leader and you could probably do with replacing. This leads to a lot of 'aggression' problems, which are not truly aggression at all but only the dog's assertion of his natural right to challenge your pack leadership!

- When you let your dog pull on the lead, he is leading the way, being your leader. This is often interpreted as surplus energy and exuberance, but it may also contain elements of dominance.

- The dog that comes up to you with a stick or a ball in his mouth and drops it at your feet, 'asking' you to throw it for him, is actually showing dominance to you by initiating play. When you automatically take the item and throw it for him, you are not only being trained by the dog to follow **his** orders – you are compounding the dog's notion that he is pack leader.

- When you give your dog his meals before you have yours, you are telling him he's the boss. Leader eats first, underlings get the scraps. That is the dog's instinct.

- The dog that guards the home excessively and growls at your visitors is taking on the role of pack leader, deciding over your head who is a threat to the pack and who is not.

It is hardly surprising that when you start reprimanding and trying to control a dog who believes you to be his social inferior, the dog becomes confused and problems are bound to follow. After all, such mixed messages would **never** occur within the dog pack.

Here's another classic example: if we allow the dog to go through doorways before us all the time, and then shut him in a room while we are out, in the dog's mind we are again giving very conflicting signals. Firstly, we need to be aware that in the wild pack, the leader always precedes the others into narrow spaces such as the entrance to the den. Thus, by allowing him to go first, granting him the freedom of the house and the right to come and go without being asked, we give the dog the impression that his standing within the pack is very high. But then suddenly, he finds himself closed in a room, his freedom round the den severely curtailed while we are out hunting and scouting and generally taking over leadership roles. What? How dare we? Nobody treats the pack leader that way!

And what's he going to do in the face of this outrageous cheek? Very possibly, wreck the place in indignation! In the dog's mind, his protest is quite justified. (Much destructive behaviour, in both the home and car, starts this way.) And so, we have another instance of 'naughty dog' behaviour entirely triggered by ourselves. And, like all forms of trouble, it's much easier to get into than out of!

The above examples are only a small selection of ways in which our ignorance of the dog's mind and values can create all sorts of problems. It is very important to learn what makes the dog's mind tick, and how we are regarded by them.

Unfortunately, the emotional healing approach to using flower remedies for dogs tends not to take much, or any, of this into account. That may work fine in simple cases of fright and trauma, as we've said, but in many other cases there needs to be understanding of the underlying psychology of the dog. In focusing on **emotions** rather than **psychology** – the result itself rather than the culture clash that produced it – this approach runs the risk of oversimplifying animal treatment and drawing misleading conclusions.

In a case like the above where the dog acted destructively after being shut in a room, someone going by the pure emotional healing approach may have decided upon the flower remedies **Red Chestnut** for a dog desperately missing its owner, or **Honeysuckle** or **Mimulus** for feeling lost, anxious and lonely, maybe **Cherry Plum** for the pressing anxiety that caused the dog to 'lose control' and tear up the furniture. But they would be blind to the real cause of the problem, shooting in the dark and quite unable to get the results that could be gained by someone who sees the case 'through the dog's eyes'. The wiser prescriber, better versed in canine psychology, would be able to offer the dog a flower remedy such as **Vine** or **Holly**, which works for states of dominance and annoyance, or better still **Willow** or **Dagger Hakea** which are very effective for behaviour based on resentment. These would take the edge off the dog's bad mood, alleviating the problem in the immediate term, and mellow him a bit. **But the flower remedies alone would not cure the**

problem entirely, or lastingly, as to a large extent the dog in this case is behaving quite normally, merely responding in dog fashion to a set of circumstances imposed on him. The effects of the essence would then be backed up and reinforced by a gentle but effective programme of 'dog demotion' (i.e. to demote the dog's standing within the pack, removing the chances of future repetitions of the problem – the programme we recommend is given at the back of the book).

This dual approach, linking together the benefits of the flower remedies with the knowledge afforded us by conventional training and psychology, is often far more effective than either on its own.

To sum up, we can say that while many conventional trainers regard dogs too much as dogs and not enough as people, many emotional healers regard dogs too much as people and not enough as dogs!

The solution, as with all things, is to achieve balance.

Towards a holistic approach to canine behaviour

The word 'holistic' means **taking the whole into account or taking a wider perspective**. It is clear that both the traditional training/behavioural psychology approach and the more esoteric emotional healing approach, have much to offer on their own. But a holistic marriage of the two vastly increases the possibilities for effective, lasting healing. There is so much more that can be done this way, tackling both simple, emotion-based cases, and complex, psychology-based cases, taking in the entire spectrum of mental, emotional, and behavioural problems of all sorts. The result is a happier dog, better integrated into his human pack, getting the right feedback from his owner; overall, a level of human/canine harmony that is otherwise all but impossible to achieve in many cases.

The purpose of this book is to continue and expand on the work of the world's flower essence pioneers – Bach, Kaminski, Katz, White and Johnson among many others – and canine behaviour visionaries like John Fisher, by helping to create a new, gentle, holistic approach to treating dog behaviour that marries the wonderful natural therapy of the flower remedies to elements of canine psychology and humanitarian training.

The book attempts to make the holistic approach as practical as possible. For any given problem, we offer flower remedy suggestions based on our own experience and back up those suggestions, where applicable, with tips on how simple training techniques can be used in conjunction with the flower remedies to help reverse or prevent the undesirable behaviour. Owners who are unsure of their dog's reasons for behaving one way or another may derive fresh insight from this approach.

The book is mainly a reference work, designed for fast access to advice on whatever problem you are experiencing with your dog. It can be dipped in and out of, or it can be read sequentially from cover to cover to gain a fuller overall understanding of what flower remedies can offer in the field of dog behaviour. But whichever way you use it, we hope and trust that it will help you, as a dog lover, to discover how your 'dogs misbehaving' can be helped in the gentlest, most effective way, enriching your relationship with one another forever.

3 Flower Remedies Covered in this Book

1. Animal Rescue (Alaska)
A special combination remedy, developed over ten years of research, to help with many forms of animal rehabilitation.

2. Arnica (California)
Mental and emotional aftermath of shock, accident, trauma, surgery, or illness.

3. Bleeding Heart (California)
Over-attachment to the owner; dogs that whinge and groan when their owner is out of sight.

4. Centaury (Bach)
Difficulty asserting one's leadership (for the owner).

5. Chamomile (California)
Dogs that bark; stomach problems caused by emotional upset.

6. Cherry Plum (Bach)
Loss of self-control; aggression caused by illness or pain; abnormal behaviours; maddening itches leading to self-mutilation.

7. Chestnut Bud (Bach)
Failure to digest lessons; repeated cycles of errors despite knowing otherwise.

8. Chicory (Bach)
Attention-seeking, manipulative behaviour; eating disorders.

9. Cosmos (California)
To encourage inter-species communication for better bonds, training, etc.

10. Cow Parsnip (Alaska)
Helping to adapt to new surroundings.

11. Dagger Hakea (Australia)
Resentful, bearing grudges, bitter and smouldering.

12. Dill (California)
Upset linked to change, travel, break in routine.

13. Dog Rose (Australia)
Everyday fears and lack of confidence.

14. Fringed Violet (Australia)
Shock, trauma and their after-effects in the short or long term.

15. Gorse (Bach)
Loss of vitality and the will to live; depression and despair.

16. Grey Spider Flower (Australia)
States of acute fear and terror in animals.

17. Heather (Bach)
Vociferous desire for attention.

18. Holly (Bach)
Aggressive behaviour, rage, jealousy, biting, fighting.

19. Honeysuckle (Bach)
Pining for the past, loss of interest in the present.

20. Impatiens (Bach)
Snappy, irritable behaviour, lack of tolerance; animals that are 'highly strung'.

21. Isopogon (Australia)
Controlling, domineering, ruthless.

22. Kangaroo Paw (Australia)
Impulsive, clumsy, unthinking behaviour.

23. Larch (Bach)
States of poor confidence and shyness.

24. Little Flannel Flower (Australia)
Repression of 'the child within', leading to overly serious children and grim adults.

25. Mariposa Lily (California)
Animals that are insufficiently bonded to their young; to help introduce a puppy to a surrogate mother.

26. Mimulus (Bach)
General, everyday fears and anxiety about known things.

27. Mountain Devil (Australia)
Chronic hostility and angry disposition.

28. Old Man Banksia (Australia)
Sluggishness, lack of energy and drive, apathy.

29. Pine (Bach)
For owners who feel guilty about ignoring their dogs during training.

30. Pink Yarrow (California)
Tendency for pets to pick up on and mirror the negative emotions of their owners.

31. Quaking Grass (California)
Helps a group of animals re-adjust after a change, such as a new addition to the pack/herd.

32. Red Clover (California)
Problems with animal hysteria; hysterical fear at the vet's.

33. Red Helmet Orchid (Australia)

Rebellious, hot-headed, fiery disposition.

34. Rescue Remedy (Bach)

Life-saving emergency tonic.

35. Rock Rose (Bach)

States of terror and extreme nervous agitation.

36. Saguaro (California)

Rebelliousness, resistance to authority.

37. Scleranthus (Bach)

Vacillating, swinging moods; animals that seem unsettled and keep shifting from place to place.

38. Self Heal (California)

Aids in stimulating inner curative forces and will to live.

39. Snapdragon (California)

Aggression with tendency to bite.

40. Soapberry (Alaska)

Fear, including fear of other animals.

41. Soul Support (Alaska)

Emergency, trauma, accident, nervous shock.

42. Southern Cross (Australia)

Bitter, self-pitying, 'victim' mentality.

43. Star of Bethlehem (Bach)

For any present or past state of trauma, abuse, shock, unhappiness, loss.

44. Tiger Lily (California)

Aggression in dogs and cats.

45. Tundra Twayblade (Alaska)

After-effects of shock, trauma and abuse in animals.

46. Vervain (Bach)

Over-enthusiasm and hyperactivity.

47. Vine (Bach)

Tyrannical, overwhelmingly dominant traits; to assist in 'demotion' of spoilt/aggressive dog.

48. Walnut (Bach)

Emotional upset linked to change, such as moving home or changing owner.

49. Wild Rose (Bach)

Loss of interest in life, emotional 'shutdown', listlessness and apathy, often following period of abuse or abnormal stress.

50. Willow (Bach)

Feeling mistreated; resentful and bitter, harbouring hostile thoughts.

4 Solving Problems

Abuse

It's an unfortunate fact that many dogs suffer at the hands of unsuitable, often cruel owners. They can be psychologically marked for life, much like people, after periods of mistreatment. Many dogs that are rescued after being abandoned may have spent time in puppy farms or other terrible environments, or may have spent a long time wandering about in fear, possibly in pain. Other dogs may have been severely beaten in their lives and flinch whenever you stand up or raise a hand.

Before you can earn such a dog's love and companionship, you have to enable him to overcome whatever traumas lie in his past. This will often have to be done before you can even think about starting any kind of training programme with the dog.

FLOWER REMEDY SUGGESTIONS

Star of Bethlehem
For releasing the stress of old shocks, traumas, bad experiences. An extremely important remedy for all animals with a 'past'.

Tundra Twayblade
After-effects of shock and trauma in animals; helps to relieve the anxiety and fearful demeanour that may remain after an animal has been badly treated.

Wild Rose
Often indicated when an abused animal has gone into a state of hopeless despondency, as though it no longer cared what happened to it. The remedy restores vitality.

Mimulus
Helps with chronic states of fearfulness that could be sparked day-to-day by any little thing, often a feature of animals with a history of abuse.

Cherry Plum
When pain, cruelty and the horrors of abuse are fresh in an animal's mind and it is liable to lash out in defence at anyone trying to help.

Animal Rescue Formula
This newly launched formula is already proving itself highly effective at healing the after-effects of abuse and trauma.

Adolescence

Adolescence usually starts at around six months and may last up to the age of eighteen months, depending on the breed of dog. This is a time when dogs often become rather unruly and hard

'It's just a phase he's going through.'

to handle and train, when many owners give up on them and have them re-homed. For the dog, it's a period of discovery, developing maturity and rebellion, much like the teen years of a human. He may not be as willing to listen to you as he was in puppyhood, and he may appear more interested in the world about him than he is in you. Accept this as normal, and be patient and understanding as the phase works itself out. Continue with your training, a little each day, always sure to end on 'a good note', even if it means you've only done five minutes.

Adolescence is, obviously, also linked to the onset of sexual maturity, and both males and females will undergo fluctuations of hormone levels with quite noticeable changes in mood and behaviour. This is also quite normal as long as it is within reasonable limits – consult your vet if in any doubt.

FLOWER REMEDY SUGGESTIONS

Chestnut Bud
For impulsiveness, youthful over-exuberance and tending not to absorb the lessons of training and discipline; dogs that may seem a bit 'scatty' and even stupid, making the same mistake over and over, despite knowing not to.

Saguaro
This remedy is very useful for adolescence and young adulthood, dealing with impulsiveness and tendencies towards rebelliousness.

Red Helmet Orchid
This is similar to Saguaro, but with even more emphasis on rebelliousness and problems accepting authority – a potentially dominant 'young upstart' can be nipped in the bud before he becomes a problem.

Walnut/Scleranthus
These will both help to deal with changing hormonal levels as the dog grows up, and temperament fluctuation (especially in bitches).

Impatiens

This is more for the owner than the dog. Adolescent dogs will frequently try our patience to the limit; however, we should not lose our temper with them as we stand to undo all the good work we've done so far. This remedy helps us to keep our cool and not get annoyed.

Aggression

Aggression is a common problem with dogs and a cause for concern, not least because a dog can inflict terrible injuries when enraged. It is important to realise that there is no one single cure for aggression, either in terms of flower remedies or in terms of training. Aggression may have various triggers, but more often than not it will be down to fearfulness, where a dog sees no option other than to stand and defend itself; or dominance, where a dog feels the use of force is necessary to maintain, defend, or raise, its social standing within the pack.

Fearful aggression is commonly seen in dogs that have not been properly socialised, or that have been through a bad time. For hints, tips and flower remedy suggestions, *see* **Abuse**, **Biting**, and **Socialisation**.

In cases of **dominant aggression**, hostilities break out when there is some doubt as to which pack member has the ability, or right, to be in control. In other words, dogs use aggression as a tool for sorting out who's in charge.

Aggression between two dogs may just be a natural process of establishing, or maintaining, hierarchy. For instance, we have an ageing German Shepherd and a young, strapping Rottweiler. The GSD has always been leader; but, as she begins to weaken, it is inevitable that the Rottie will see his chance to take over. More and more, he is getting the better of her. When they squabble and she drives him away, the *status quo* is being maintained; when he manages to take a bone or a ball from her, it is a glimpse of his future status as Alpha dog. She knows she is in decline, and losing her grip on her power. Perhaps this is what people mean when they say: 'Nature is cruel.' Really, it is neither a good nor a bad thing – it's just a normal, neutral process for furthering survival. We do not interfere with their contests for leadership, and one way or another it will be resolved between them.

On the other hand, there are times when dog-on-dog aggression is not normal, but has been artificially provoked by us when we interfere in matters of pack hierarchy. In those cases, one may treat using flower remedies. *See* **Jealousy**.

A dog's reasons for showing aggression of the dominant, socially 'ambitious' kind towards people is not much different from that towards other dogs. The motive is the same, i.e. to contest rank. For more information on how to solve these problems, *see* **Biting (Dominance-biting)** and **Dominance**.

Attention-seeking

This is another common problem that plagues dog owners. It can become a nightmare when your dog just will not leave you alone, or will not allow you to leave him alone, which is more to the point.

Attention-seeking is really about control. Dogs are very good at getting our attention, and they have many ways of making sure they get it whenever they want it. We can become like an 'affection-tap' that they can turn on whenever they please.

It often begins with our feeling flattered when our puppy or new dog starts coming over to us to 'say hello' in the form of a lick, or a nudge, as we sit reading or watching TV. We will automatically respond with a pat on the head and a 'good dog', and maybe sometimes a treat, too. In the dog's mind, the way to get affection and treats is to come up and just ask for them – works every time! Before we know it, the dog is getting more and more insistent and constantly interrupting us for attention. Furthermore, as the situation escalates, many dogs start to dislike it when your attention is paid to anyone or anything else. Examples are the telephone, the TV, books and newspapers, other people (especially visitors), and, in one extreme case that we dealt with, the view from the window! (*See* **Case Histories:**

'It's not as interesting as me.'

Toby.) To get attention, dogs may whinge, bark, scratch at your leg, place a paw on you, stare pathetically at you, press themselves up against you, or even – larger breeds find this easier – reverse into you and sit on your knee. This is a great party trick that is all too often encouraged by the laughter of visitors (*see* **Visitors**).

There are times, of course, when your dog may have a very valid reason for trying to flag down your attention. He may be hungry, thirsty, or want to relieve himself. Sometimes it's hard, especially with very young or old dogs that may be somewhat incontinent, to know when whingeing at 3 a.m. is attention-seeking or an urgent cry to go out for a piddle. Make sure such needs are attended to before concluding that the behaviour is attention-seeking.

One way of preventing attention-seeking behaviour is not to give your dog attention on demand. This is not to say that you should not show him affection. What you should do when he comes up is ignore him, send him away gently but firmly, showing him that you're not interested in what he has to offer right now. Then, ten or fifteen minutes later, making sure he has totally given up trying to get your attention, call him over and make a fuss of him. What you are doing here is taking control back from the dog. He no longer decides when he gets the attention – **you** decide.

If you have a problem with an attention-seeking dog and you decide to start ignoring him and controlling the situation by making the first moves as a rule, you may find the dog resisting you and perhaps even sulking. Barking dogs may bark louder and longer, whining dogs will wail like banshees, and staring dogs will bore into you with laser-beam eyes. The reason is that they suddenly find their method for getting your attention is no longer working, and so they are 'turning up the volume'. Give in at that point, and you're in trouble. The problem will never resolve itself. Flower remedies can also help with this transition.

FLOWER REMEDY SUGGESTIONS

Chicory
This is one we would often go for first in these cases. It is for animals and people that are attention-seeking, often petulant and sulky when refused, and very full of self-pity. Think 'spoilt brat' and you'll be on the right track.

When the controlling behaviour is very entrenched indeed and bordering on a tyrannical grip on the household, one could opt for **Vine**. **Isopogon** is a very good alternative, from the Australian Bush family.

Heather
This would also be helpful for reducing a dog's cloying, attention-seeking ways, especially if he is very vocal with it.

Chamomile
Quite similar to **Heather** and of use for dogs that bark a lot to remind you they are there.

And for the owner, trying to ignore the dog:

Impatiens
To reduce the stress of having to wait while the dog's methods of attention-grabbing diminish. If you have allowed the situation to become extreme, the constant barking and howling can drive you crazy.

Centaury
This helps people assert their willpower and not give in to the dog's controlling behaviour.

Pine
Owners often feel guilty about ignoring their dog. In fact this is one of the reasons why training programmes sometimes break down. There is often someone in the family (usually someone on the outside who doesn't have to put up with the dog all the time) who thinks the new regime set out by the trainer is cruel or unfair. Don't listen to them! This remedy helps alleviate the sense of guilt and helps you understand when guilt is unfounded.

Biting (and mouthing)

There are two basic forms of biting: fear-biting and dominance-biting – *see below*.

Fear-biting

This is the more common reason for biting, but often develops into dominance if unchecked. In any case, you will want to gain the dog's trust in order to prevent getting bitten. *See* **Abuse**; **Fears and Anxieties**; **Socialisation**.

FLOWER REMEDY SUGGESTIONS

Rock Rose
For outright terror and panicky states.

Cherry Plum
For fear that makes an animal lash out in desperation.

Snapdragon
For animals that are quick to bite in pressing situations.

Red Clover
For the tendency to hysteria and uncontrollable behaviour.

Dominance-biting

Biting of this sort is yet another example of poor upbringing, and mouthing is the thin end of the wedge. It may be fun to play with a puppy that mouths and chews at our fingers, but we are teaching the dog behaviour that may later turn to outright dominance. However playfully they may do it, dogs that grab, bite, or mouth parts of our anatomy, are really 'trying it on', giving us a display of strength that says: '*Watch it – I can hurt you if I want to.*'

Lots of people enjoy play-fighting with their dogs (especially men with big dogs), and what frequently happens is that as the

level of roughness escalates, the dog starts to use his mouth to grab at arms or hands – at which point the human backs down, intimidated and trying to defuse the situation. It's a case of Dog 1, Human 0! The dog will remember how to win the game, and next time he'll mouth and bite more readily and with more force. This will apply not just to the owner but to any person the dog wants to intimidate. When a dog has thus been taught to bite, this will almost always cause great problems later on, and many dogs are unjustly destroyed each year under the Dangerous Dogs Act (UK) and equivalent laws in other countries, for having been allowed to bite people. Most of the fault lies with their owners. *See* **Dominance**.

FLOWER REMEDY SUGGESTIONS

Isopogon, Vine
For domineering, controlling tendencies.

Holly, Snapdragon, Tiger Lily
For fierceness, anger and aggressiveness.

TRAINING TIPS

Let puppies take out their biting instincts on their toys rather than a human hand. Always preserve your status as pack leader. Nip any dominant traits in the bud. Don't play rough with your dog, as no person can bite harder than a dog!

Bonding

There is a theory amongst some trainers that if a dog is not socialised with humans before it is fourteen weeks old, it will be very hard and maybe even impossible for an owner to form much of a bond with it.

In our experience, we have not found this to be entirely the case, and it is a shame that such a belief is so widespread as it often deters people from adopting perfectly good dogs from sanctuaries and rescue centres. It is true that the sooner you start, the better. But if you go out of your way to spend time with an older dog, it can be made to bond with you very well. As an example, many people take in dogs from rescue centres that have been stuck in kennels and puppy farms during their most formative phase of life and never been bonded with humans before, and yet they form wonderful mutual attachments with them. (In some cases, there will be traumas in the dog's past that may block bonding, but these can be treated very effectively – *see* **Abuse**.) It's also wrong to think that getting a dog younger than fourteen weeks guarantees some kind of magic bond-forming. Some people get puppies with the intention of using them as guard dogs: the dog lives outside and sleeps in a kennel, and its only contact with the owners is when they feed it. When the dog grows up, these people are often surprised when it feels no great desire to guard them. Seen from the dog's point of view, why should it? Don't you have to be **worth** guarding?

If you want to form a good bond with your dog, you can –

'I'm all right, Jack.'

regardless of its age. Invest in the process by spending time playing games together in the garden or park. (Games are an important part of bonding with your dog. But do be careful of games like 'tug-of-war', in case of creating a possessiveness problem – *see* **Possessiveness**.)

Having fun with your dog will create a strong bond that will make any later training easier, less stressful for both of you, and more effective.

FLOWER REMEDY SUGGESTIONS

Cosmos
To help promote strong one-to-one relationships. This is very useful indeed for both animal and owner/trainer to take together.

Mariposa Lily
This remedy assists in general bonding between different animals or people. It helps the animal become more receptive to its human keeper's love. It is also used for mother and child bonding, and is good for cases of bitches that show little interest in their pups.

Quaking Grass
Enhances a sense of community, harmony, togetherness and teamwork. It could be taken by everyone within the pack, human or animal.

Car problems

There are many different problems that can occur with dogs in cars. Here are a few ideas.

FLOWER REMEDY SUGGESTIONS

Dill
For the overwhelming feelings and/or confusion that may come

with being driven in the car.

Mimulus
For general fears and anxieties; dogs that whine and tremble in the car, and so on.

Grey Spider Flower/Rock Rose
For dogs whose fear borders more on terror, with panicky behaviour and perhaps scrabbling to get out.

Star of Bethlehem
If fears and travelling problems have come about because of a past trauma associated with the car (such as through being a passenger involved in an accident), this will help release the after-effects and the memory of the incident.

Scleranthus
Often handy for cases of motion sickness. This is one of the few instances where flower remedies actually come in handy for physical problems. *See also* the homoeopathic remedies below.

Soul Support
This brilliant Alaskan combination remedy is highly effective as a general calmer. We have seen dogs that were normally quite distressed in the car, fall asleep within a few short minutes of setting off and lie there like rag dolls, blissfully carefree and relaxed.

HOMOEOPATHIC REMEDIES

Cocculus, **Ipecacuanha**, **Tabacum** and **Gelsemium** are all useful for travel problems like fear, nervous vomiting in the car and motion sickness. Consult your homoeopathic vet.

TRAINING TIPS

If your dog is scared of going in the car, try leaving the back open when it's parked in the drive and letting him spend some time in there with a nice bone to chew on. (**Be extremely** careful in hot weather, when cars heat up very quickly - often with fatal results.) Nervous dogs can be made to associate the car with pleasant things. Build this up gradually, with short trips and rewards for calm behaviour. The sooner your dog gets used to travelling, the better.

It's obviously normal to take your new puppy home in your arms, sitting in the front passenger seat. But as quickly as possible afterwards, make your dog understand that his place in the car is the back. The best car for dogs is an estate car or large hatchback, where the dog can have his place behind the seats intended for humans. Too many dogs develop dominance traits in the car because of being allowed to sit where they please. Many large dogs are even allowed to occupy the front, which is not only dangerous but also is giving the dog the impression that he's high up in the pack. This may, among other things, bring about a destructiveness problem. (*See* **Dominance**, and **Case Histories** later in the book.)

Destructiveness

See **Dominance**, **Separation anxiety**.

Dominance

This is a very common cause of problems experienced by owners with their dogs. Allowing your dog to dominate you is the root of many evils. Often it isn't recognised until the situation is far advanced (see the case study of Buster), and it can take a lot of work to get things straightened out.

What is dominance? Simply put, dominance is the influence necessary for a leader to maintain control of his pack. Dogs dominate one another all the time, as a way of establishing and maintaining pack hierarchy, for optimum pack functioning and, even though they are no longer in the wild, **survival**. It is perfectly normal to see a pack leader occasionally reinforce his or her position by placing a paw across the neck or shoulder of

Learn to understand your dog's body language.

an underling. The pack understands this very clear body-language and there are no misunderstandings amongst dogs.

When we bring dogs into our home, we should be very clear in understanding that **they consider us as members of their pack**. They do not, as is often thought, automatically regard us as their leaders. Why should they? Just because we are humans? It would be sheer arrogance to assume this. Just like any other member of the pack, we have to earn our status.

Unfortunately, people are often not very well clued-up on doggy values and body language, and often take overt signs of dominance for affection. A famous dog trainer told us a story about a woman who turned up for a consultation with a huge Pyrenean Mountain Dog that she claimed was being 'naughty'. The trainer took one look at this obviously dominant dog and explained to her that the dog had no respect for her. The woman was outraged: 'How *dare* you! Of *course* my dog loves me – why, all the way here in the car, he sat next to me with his paw on my shoulder!'

Few people realise the many subtle ways in which they let their dogs dominate them. In many cases you can get away with it, and no problems occur, but it depends on the character of the dog. In many other cases, as the dog becomes more dominant he starts to become more controlling, and his behaviour may become very problematic indeed. The dominant dog that thinks he has a right to rule the household will make his own decisions. He won't obey commands. He will refuse to do anything he doesn't feel like doing. He will challenge your authority if you disagree with his behaviour. He will demand the privileges of the pack leader, which include sleeping where he likes (your bed is a great symbol of rank), access to all parts of the house, and the right to be the centre of attention at all times. Failure to comply with any of the above may result in aggressive or destructive behaviour, which is the only option open to the dog to reassert w**hat he had been led to believe** was his leadership right. Seen in this light, you can't really blame him!

Dog trainers repeat it like a mantra: **keep your dog's rank low within the human pack**. In fact, the dog should be right at the bottom. This is not cruel or harsh in any way. Most dogs actually thrive on knowing clearly where they stand within their pack.

Keeping your dog low in rank is achieved by giving him consistent messages. These can become second nature in your daily routine, so that you can 'train' him quite effortlessly. Flower remedies will back this up where necessary, and are very useful with the occasional dog that gets above itself or does not like being dictated to.

FLOWER REMEDY SUGGESTIONS

Vine/Isopogon

These are the key remedies for addressing outright dominance. Dogs that let their dominating urge run away with them and become tyrannical and ruthless become much more relaxed and 'mellow' on these remedies. In less extreme cases they have been seen to bring the dog's behaviour down to a level that owners find acceptable without the need for any training. But to really get 'under' a more serious dominance problem the remedies will probably need to be reinforced with a programme to reverse the dog's rank in the pack (*see* the **Dog Demotion** programme at the end of the book).

Dagger Hakea/Willow

These are the ones to go for when your dog resents being subject to your authority. You will see how the dog becomes moody and sullen when you give him a command. When owners start introducing a training programme in response to some behaviour problem, some headstrong dogs may resist it quite forcefully to begin with. Sometimes their bad behaviour may get even worse for a short time! Trainers call this the 'extinction burst'. Giving **Dagger Hakea** or **Willow** to the dog at this point, or as you start the training, will often allow the new regime to get going more smoothly.

Saguaro

This is especially useful in cases of adolescent dogs that are beginning to challenge your authority as they approach maturity. It works well with the above two.

If your dog combines dominant traits with a tendency to get hostile or aggressive with you, as well as addressing the underlying dominance (cause), you will be well advised to address the aggressive behaviour too (effect). There are several flower remedies available to help with this: **Holly**, **Tiger Lily**, **Snapdragon**, **Mountain Devil**, are all useful and any of them would definitely help reduce the level of aggression.

Remember that in some cases, extreme aggression may be the result of an illness. If your dog suddenly becomes hostile for no apparent reason, this may be what is called 'idiopathic aggression' and means you need to call the vet to diagnose whatever underlying problem is sparking the behaviour. **So, if in doubt, call your vet right away.**

Emergencies

See **Traumas and emergencies.**

Fears and anxieties

There are times when our dogs can cause us problems, not by 'misbehaving' as such, but because of pure emotions. Fear and anxiety are things we all know, and we can understand subjectively, without having to go into dog psychology, how deeply fear can affect our pets.

Fear comes in many different shapes and sizes. There are the frights that accompany traumatic events such as being attacked by another dog, or being in an accident (*see* **Traumas and emergencies**). Long-term fearfulness may result from these incidents. Training can help, and has various techniques at its disposal. But flower remedies get right to the root of the matter, repairing at an energy level the emotional damage that is causing the fear in the first place. Let's look at some flower remedies that could help with the various types of fear a dog may experience.

FLOWER REMEDY SUGGESTIONS

Mimulus and Dog Rose

These two remedies basically address similar types of fear – namely everyday, ongoing mild fears and anxieties. They are not generally used for acute panic or very extreme fear or terror. People and animals that have developed mild phobias will benefit from these remedies. They will also help in cases where traits of fearfulness, nervousness, shyness, and lack of confidence are part of an animal or person's character. So a dog that lacked confidence in a show situation, despite never having had an actual fright in that situation, would benefit. You would find that its confidence improved generally, enhancing its interaction with its environment in all sorts of ways.

Fear, like all dis-ease, can have two sides to it. A chronic, ongoing state of fear may be born out of an earlier shock or fright. The **Mimulus/Dog Rose** would address the **result** of the fright, while you could simultaneously address the **cause** of the fright with a deeper-acting remedy like **Star of Bethlehem**, **Fringed Violet** or **Tundra Twayblade**, all of which can heal the damage caused by trauma that happened years ago.

Rock Rose and Grey Spider Flower
Again, we have two remedies here that are quite similar in application. These would be suitable for more extreme fears, terror, panic. It is quite easy to tell the level of fear in a dog. A very scared dog will tremble, cower, perhaps roll over in submission, perhaps wet itself, or just freeze to a standstill in absolute terror as it gives in to imminent death. In these cases, the remedies for real terror and very strong acute fear are called for. Again, as above, if the fear is rooted in a past trauma, it is desirable to address the cause as well as the effect.

Soapberry
If your dog has been attacked by another dog, and become

Dogs can often pick up on their owner's fears.

compulsively fearful of other dogs as a result, this remedy would be worth adding to the above. We have also found that it is especially useful for people who are afraid of animals. This is more an irrational, phobic fear than the more practical fear of being hurt, which is the more usual reason for fearing a dog.

Red Clover

This is very helpful for the kind of struggling fear and desire to escape that many animals display when taken to the vet's. It would probably also do for people in the dentist's chair!

Rescue Remedy/Soul Support

These combinations will be of great value in calming acute states of fear, and are always worth carrying in your pocket 'just in case'.

A word on 'unknown' fears: often a dog can seem frightened for no apparent reason; a fearful state may just 'come on', seemingly without cause. In our experience, there is **always** a cause. The fear may be 'unknown' to us, but it is not unknown to the dog. The dog has his own reason for being afraid. Something must have frightened him, or reminded him of a frightening incident in his past. Therefore, in cases of so-called 'unknown' fear, rather than go for the Bach remedy **Aspen** as is often done, it is advisable to treat for trauma – even if you cannot identify the trauma. If you are wrong, no harm can be done; the remedy just fails to act.

NOTE

Dogs can also pick up on their owner's fears and anxieties. We have often found that dogs that are afraid of fireworks have owners who are afraid of fireworks. In these cases, it may be that treating the dog alone does not achieve a great result. (*See* **Case Histories: Fred**.) But if the owner is causing the dog to be fearful, it is their responsibility to get their own emotional difficulty sorted out.

Remedies that help to protect a dog from the negative atmosphere created by an owner's own fears, hang-ups and emotional problems are **Pink Yarrow** and **Walnut**

Fouling

House-training is quite a complex subject and pretty much lies outside the scope of this book. Obviously, if you leave your puppy all night without going outside to relieve himself, no flower remedy ever devised is going to make him hold it in so that you don't have to face a cleaning-up job in the morning.

There are cases, though, when fouling is related to dominant behaviour, separation anxiety or frights and traumas. You will need to look up each of those sections, if relevant, and decide on the best options.

Hyperactivity

This is very likely to be down to dietary matters, lack of exercise leading to boredom, or a lack of control on the part of the owner. A dog that has had good basic training, gets adequate exercise and a decent natural diet will tend not to act hyperactively.

Before you get a dog, consider your own lifestyle and how it will fit in with the requirements of different sizes and breeds of dog. A Border Collie, bred for running over wide spaces in the great outdoors, will often get bored and become a problem in a small flat in the city.

The following remedies can help in some cases, but it is most important to isolate and rectify the real cause of this much-misunderstood problem.

FLOWER REMEDY SUGGESTIONS

Vervain
Useful for excessive exuberance and the tendency to use up too much energy.

Chestnut Bud
For youthful impulsiveness.

Impatiens
Nervous and highly strung animals that rush around.

Chamomile
For wild behaviour, hysteria, nervous barking.

Rescue Remedy/Soul Support
General calmers.

Jealousy

Dogs can often become jealous, of each other, when there is more than one in a household, or of anyone who appears to be threatening or detracting from their position. As we've said, dogs are very conscious of their pack hierarchy. So if we fail to respect that, we might unknowingly interfere in their social politics. In fact, dogs in the wild seldom experience jealousy since their social system functions a lot better without human meddling!

A classic example is when we have a favourite dog out of our two, three, or more, and we grant it favours such as treats, walks, and cuddles which the others don't get; maybe we tend to feed it first, or let it sit on our knee when the others must stay on the floor. The others will observe all this, and may be very resentful, giving rise to skirmishes and fights. Another example is when we introduce a new dog or pup. We should always let the top dog (that is, the leader amongst the dogs – not the overall pack leader, which should be **YOU**), have precedence.

There have been a small number of tragic cases where dogs have attacked and even killed their owners' baby. This is almost certainly a result of jealousy, when the dog is suddenly ignored in favour of a newcomer. In the dog's mind, disposing of the baby will restore the old, preferable situation. These tragedies also highlight the potential risks of allowing a dog to become dominant within the pack, where it is liable to act on its own decisions. **This book does most certainly NOT advocate that flower remedies alone will protect your baby from a jealous** dog. Never leave your baby alone and unsupervised with any dog, regardless of breed or size.

NOTE It is not just dogs that pose a threat to babies: there have been many cases of older children trying to harm their new siblings out of jealousy. Would you leave your baby in the care of a three year-old?

FLOWER REMEDY SUGGESTIONS

Quaking Grass
When a new animal has been introduced to the family, this remedy can be given to the newcomer and any existing animals, to help them all adjust to each other and the new situation.

Holly
A very important remedy to consider if there is any sign that one animal is jealous of another, or jealous of a child or baby. It reduces the aggressive urge, bringing greater calm and tolerance.

Willow/Dagger Hakea
Useful when a dog appears resentful, or unusually moody, after someone else in the pack has had preferential treatment. Also effective (with **Holly**) for dogs tending to be dominant and that don't take kindly to your reprimands.

Southern Cross
This is a remedy for resentment combined with the 'victim mentality'. A dog that appears ill, or will not eat, and has no medical reason for acting this way, might just possibly be doing so as a protest against some act of unfairness that it perceives you have committed against it. This might sound outlandish, but we have experience of at least two dogs that have benefited from this remedy for this very problem. *See also* **Attention seeking**.

Possessiveness

Dogs that become possessive of such things as toys, balls, sticks, and items of furniture such as beds, sofas, armchairs, are generally being dominant. (*See* **Dominance** for more information on this common aspect of dog behaviour.)

Owners often encourage this behaviour without realising it, by playing 'tug-of-war' games with their puppies or grown dogs and letting them win all the time. There is nothing wrong with playing these games as long as you win more often than the dog, always win the last game of the day and always keep possession of the 'trophy' when the game is over. All this will prevent the dog from thinking that he is stronger than you and has a right to keep hold of the items. **Remember – it may just seem like an old rubber ball or bit of rope to you, but to the dog it is a very important symbol of rank and power, just like a crown or a sceptre.**

If you think you have a problem with a possessive dog, please refer to the **Dominance** section for advice on how to control it. Very pushy pups that snap at your fingers when you try to take their toys away may well turn into future possessive adults, and this can be nipped in the bud in the same way.

Pulling on the lead

Many dogs pull on the lead. In fact, when you walk through the street it's amazing how many dogs you see taking their owners for a walk, rather than the other way round.

A lot of dogs do it because they are excited about where they are headed. There will be a reward at the end of it, such as running free with the other dogs in the park. The dog is displaying happiness, which is nice – but at the same time, the owner risks being pulled off their feet, being injured, or losing control of their dog, which may then disappear under the wheels of a passing bus. So it's a good idea to try to control this kind of exuberance. (*See* **Hyperactivity**.)

FLOWER REMEDY SUGGESTIONS

Vervain
For over-enthusiasm.

Impatiens
As the name would imply, for 'I can't wait to get there!'

Chestnut Bud
For impulsive behaviour and repeated mistakes. (We did once give some of this to a dog that used to half-strangle himself by straining like mad on the lead. He did this every time, and every time he would choke and splutter for a long while afterwards as though he could hardly breathe. This dog tended to start every day 'from scratch', never learning from his mistakes, though in other ways he was a good dog. After **Chestnut Bud**, the straining became much less pronounced, as did his repeated cycles of errors generally.)

Other dogs pull on the lead for reasons of dominance. The pack leader does exactly that: he leads. A dog that respects his owner's superior rank will stay at heel.

The dog that pulls exuberantly, out of excitement, will not tend to pull when heading for home – playtime's over, whereas the dog that pulls for reasons of dominance will pull all the time, and combine this with barking aggressively at other dogs he sees. If your dog is like this, it may not be just the lead-pulling which needs attention – you should be re-examining his whole relationship with you. *See* **Dominance** for flower remedy ideas and training tips.

Resentment

This may be connected to jealousy, such as when an owner gives preferential treatment to a favourite dog and unwittingly offends another who is higher up in the pack. (*See* **Jealousy**.)

'Resentment.'

Separation anxiety

This is a large category which covers many different problems. Basically, it refers to dogs that for one reason or another can't stand being left alone. Sometimes dogs that have not been used to solitude are suddenly put into situations where they are alone for long periods: for instance when owners take a new job outside the home. Other examples occur when dogs have become too used to following their owners around the house and will not tolerate being left on their own. There is often an element of dominance to this.

Problems such as howling, destructiveness (e.g. chewing, scratching, digging), fouling, may all occur with separation anxiety. Flower remedies can help, but the onus is also on the owner to bring up the dog in a way that he is used to being alone and can cope with it. Don't let him always follow you from room to room; don't make a big fuss of the dog when you leave the house – act casually and the dog will not think too much of it. If the dog is bored, leave him a tasty and long-lasting snack like a marrowbone stuffed with biscuits, or a specially designed toy such as a Roll-a-Treat (a rubber ball that dispenses treats), to stimulate him mentally and pass the time. Only let the dog have the treat or toy when you are out, and take it away as soon as you come home. That way the dog will actively look forward to your leaving the house! Also, try having the radio on when you are in the house and leaving it on when you go out. The dog will derive comfort from that 'common factor'.

FLOWER REMEDY SUGGESTIONS

Honeysuckle

This is useful for many of the more 'straightforward' cases of separation anxiety, where a dog is just 'feeling the blues' because of the owner's absence. It may express this by soulful howling, a wistful expression and pawing at the bars of its enclosure. Honeysuckle is for states of 'I want things back to the way they were before', and so would also be useful if the dog were pining for an owner who will not return, possibly who has died.

Gorse

This would be called for in a dog that is truly despairing, close to grief, over the absence of a beloved owner. Again, when an owner has passed away, a dog may pine bitterly and sometimes even give up the will to live. Gorse is a great rebuilder of the vital force. Also excellent in this and many other respects is **Self-Heal**, which assists recovery of the will to live.

Bleeding Heart

We have found this effective for dogs that mope around waiting for their owners to return, becoming bored and self-pitying. Dogs in need of this remedy will tend to pour too much of themselves into their owners, and this is usually compounded by sloppy training.

Chicory

Chicory is suitable for the dog who is a little bit more 'active' in his separation anxiety. He is usually the dog who has been the centre of attention, the apple of his owner's eye, always pestering when visitors come, often barking to get attention when the owner is on the phone. When left alone, this type of dog is often a problem. There is a dominant element to this behaviour that may additionally call for real adjustments to the way the household is run. You may create a **Chicory** dog by allowing him to sleep on your bed, beg at the table, etc. Refer to the **Dog Demotion** programme at the back of the book.

Chicory will, again, help soften and moderate the dog's attention-seeking behaviour, but it is unlikely to perform the whole task on its own. The rest is up to you.

Willow/Dagger Hakea/Vine

These address the very dominant dog that feels actively resentful about being left alone. **Willow** and **Dagger Hakea** will work on the seething resentment, while **Vine** helps to reduce the dog's urge to dominate and regard itself as the all-powerful ruler of the pack. We have had some great results using these remedies (*see* **Case Histories**), but remember the importance of backing up such treatment with sound and **consistent** training/hierarchy measures (*see* **Dominance**).

Socialisation

This is a very important part of any dog's training. Dogs have to be socialised from an early age with people, other dogs, any other animals they may encounter, and with life's experiences in general. If a dog isn't properly socialised, it may grow up to be a fearful, unpredictable and sometimes aggressive animal. It is possible to socialise an older dog, but the earlier you start, the better. Many dog training books give useful tips on socialising your puppy or adult dog.

Flower remedies can help with the process of socialisation, by easing some of the blocks such as lack of confidence, fears, etc.

FLOWER REMEDY SUGGESTIONS

Mimulus/Larch

These two Bach remedies would be very useful to give to any young dog being introduced to the world, people and other animals for the first time, if it showed any signs of nervousness and shyness.

Dog Rose

This remedy more or less covers the combined benefits of **Mimulus** and **Larch**, as a 'two-in-one'. General anxiety, niggling fears (not terror), and lack of confidence are the indications for this remedy.

Cosmos

Helps encourage inter-species communication. That may sound

'Pack harmony.'

a bit other-worldly, but all it means is that it helps with introducing animals to one another, such as a cat to a dog, a dog to a horse, a puppy to children (in the dog's eyes, we also count as animals). The remedy is also good for **intra**-species relations, e.g. dog to dog, cat to cat etc.

Dill
This will help soothe the dog's reactions to first-time experiences which may be overwhelming, such as being taken out into the street for the first time and seeing hundreds of new sights: people, traffic, and so on. These experiences may be very daunting to a young pup and **Dill** will help it to assimilate all the information without suffering 'sensory overload' and getting physically and mentally exhausted, scared or confused (try giving this one along with **Dog Rose** or **Mimulus**).

Territory

See **Dominance**.

Traumas and emergencies

Traumatic events in a dog's life may include all manner of things: being attacked by another dog, or being beaten by a bad owner, or being hit by a car, suffering severe frights such as from fireworks. Other types of trauma might include losing an owner or companion (*see* **Separation Anxiety**), or moving home. Often training can do little to ease these shocking events, as they are more to do with pure **emotion** rather than **behaviour** as such.

FLOWER REMEDY SUGGESTIONS

Star of Bethlehem
This is great for all forms of shock and trauma, whether it is a recent happening or something that happened long ago. It is one of the constituents of **Rescue Remedy** (*see below*).

Fringed Violet

This is also highly effective, especially good where a dog does not seem to have recuperated one hundred per cent after a traumatic incident.

Arnica

Not to be confused with the well-known homoeopathic remedy for sprains, bruises and muscle pains, this is the essence made up from the flower of the same bush. It is very similar in operation to the above two flower remedies, used for the aftermath of shock and trauma. It is also of help in getting a dog over the traumatic hurdle of major surgery or illness.

Rescue Remedy/Soul Support

If you are looking for a combination remedy to stand by in case of accidents and emergencies, either of these would be very good indeed. **Rescue Remedy** is probably the most famous of all flower essences and is actually a combination of five Bach remedies. It has a solid reputation for all emergency situations, such as deep shock following accident or injury, collapse, severe frights, etc. **Soul Support** is its Alaskan equivalent, a combination of six flowers, two gem essences and a source of remedial spring water. In our experience and opinion **Soul Support** is quite amazing and even better than the highly effective **Rescue Remedy**. We would personally recommend this one to any owner of a dog or other animal.

(In the case of an accident where your dog has been hurt, for his sake you must remain calm and clear-headed in order to think fast and call the vet, or get the dog to the vet, as quickly as possible. We would definitely advise that for every dose of **Soul Support** you give the dog, you should take one yourself.)

A common after-effect of a traumatic incident is chronic fearfulness – *see* **Fears and anxieties**.

Less serious traumas, such as being uprooted from a loved home or having a reliable routine broken, can also nonetheless be quite upsetting to a dog. The Bach remedy **Walnut** is well known as a remedy for all the negative effects of change of different kinds.

Also useful for minor traumas involving breaks in routine or having to adapt to a new environment, would be the Californian remedy **Dill** and the Alaskan **Cow Parsnip**.

Visitors

How does your dog react when visitors come to the house?

The dog that jumps up at people in a massive display of 'affection', and the dog that growls at people when they come into your home, are actually both exhibiting different sides of the same basic attitude. You guessed it – once again, **Dominance** (*see* page 43) is the key to this behaviour.

It all boils down to who is pack leader. The leaders are the ones who decide how a visitor to the house should be greeted. The leaders also control the entrances and exits to the house. So if your dog takes it upon himself to greet visitors, either by rolling out the red carpet for them or acting as a 'bouncer' on the door, he basically is taking the lead role.

Making visitors comply with your training programme is

'The Bouncer.'

probably one of the biggest headaches for any dog owner. Even professional trainers have the same problem when their friends or relatives just will not understand that under certain circumstances a dog has to be ignored and that it is very easy to undo a lot of good work when the self-proclaimed 'dog lover' makes a big fuss of a dog when it is jumping up showing dominance. Their high-pitched cries of encouragement are guaranteed to get the dog excited, and are a great reward for bad behaviour.

Visitors must be made to realise that they are not to interfere with the way you decide to bring up your dog. In disregarding your wishes, they are doing your dog a real disfavour, and showing you a marked lack of respect into the bargain.

FLOWER REMEDY SUGGESTIONS

Centaury
A good remedy for owners to use when they have trouble asserting themselves to say: 'Now look here, would you mind not... etc, etc.'

Vervain/Chestnut Bud/Kangaroo Paw
A combination of **Vervain** and **Chestnut Bud** or **Kangaroo Paw** may also help to quieten a dog that goes mad rushing around and crashing into furniture and people's legs when visitors come, excited by the build-up of anticipation with which you are filling the air.

5 Case Histories

Cassius

Cassius, a handsome and athletic Boxer (get it?), was a successful and confident show dog until, one day, an over-enthusiastic judge handled his mouth too roughly and caused him pain. From then on, Cassius developed not only a fear of going to shows, but fearful aggression towards men in general. His owners, Eileen and Gordon, who loved him and doted on him, did not mind so much that his show career seemed to be over, but they did mind that he could no longer be taken out for proper walks. Every time he saw a man, he would cower and snarl ferociously. His relationship with Eileen remained fine – but he had come to feel suspicious of Gordon, and Gordon, in turn, didn't feel too comfortable with him in case he took a bite out of him. A happy family had been badly disrupted by one foolish error.

Cassius had been seen by a vet, who prescribed tranquillisers which simply knocked the dog out and rapidly ended up in the bin, and a trainer who tried a few techniques to no avail. The owners contacted us after reading about flower remedies in a magazine.

It was very simple to work out what Cassius needed. We gave him all Bach remedies: **Star of Bethlehem**, **Holly**, **Cherry Plum** and **Mimulus**.

Result: As the 30ml treatment bottle was coming to an end after about three weeks, the effects of the remedies began to show. First of all Cassius stopped being fearful and aggressive with Gordon, and although he still showed fear when he passed men in the street, he stopped snarling. It seemed as though the aggression 'layer' of his behaviour had been removed.

Cassius received a second bottle a week after the first one was finished. This time he was given only **Star of Bethlehem** and

Mimulus. After a week on these, Gordon and Eileen phoned to say they had just deliberately taken Cassius out for a walk into town, where he had met lots of people and been petted by male strangers. He hadn't flinched – he was quite back to normal. Gordon and Eileen wanted to know whether to go on and finish the bottle, but we said this wasn't necessary.

Cassius has never suffered any kind of relapse. When we last spoke to Gordon and Eileen a few months ago, they were seriously considering taking him back into the show ring – something they had previously more or less given up hope of.

Jessie

Jessie, a Dalmatian, had been used as a breeding bitch in a puppy farm in North Wales. She had been beaten and half-starved and was in a terrible condition when the puppy farm was uncovered and the owners prosecuted. Jessie wound up in a dog sanctuary. After her sores healed and she had filled out a bit, her looks were so stunning that many visitors to the sanctuary wanted to adopt her. The only problem was that when anyone tried to approach her, she would cower at the back of her kennel and tremble all over. She was completely submissive, and there was no hint of aggression – but prospective foster-owners were put off by her behaviour. They didn't see how they could cope with such a fearful dog.

Jessie didn't relax with anyone, not even the members of staff who cared for her each day. Before we saw her, she had been in the sanctuary for nearly two months and there had been no trace of improvement in her behaviour.

A vet whom the sanctuary dealt with knew us, and suggested that perhaps Jessie's problems could be helped. We visited the place, and saw for ourselves that the Dalmatian really did have a serious trauma problem. We wanted to do a really thorough job with her, which might take time, and asked the manager of the sanctuary how long they thought Jessie would be there. 'At least three more months, if not longer,' the manager replied, 'in the state she's in.'

We immediately prepared a 30ml treatment bottle containing

Star of Bethlehem for the effects of trauma, and **Grey Spider Flower** for extreme fear bordering on terror. We left instructions with staff on how the drops should be given (4 drops, 4 times a day in food or water), and told the manager we'd be in touch in one week to check on Jessie's progress.

We phoned back in a week as promised, and to our amazement we were told that Jessie had been rehomed two days earlier. A nice family had come and taken her away. When we asked why the dog had been taken away so much sooner than expected, the manager told us that within two days of starting on the remedies, Jessie had calmed down and stopped exhibiting fearful symptoms. She had started coming forward to the front of the kennel, sniffing at people when they approached, and after a further two days was licking people affectionately. The girls who cleaned the kennel and fed her could hardly believe it was the same dog.

We contacted the new owners to ask them if they would kindly continue with the flower remedies until the bottle was finished. A few weeks later they phoned back to say that Jessie was showing no sign of fear or anxiety whatsoever – in fact they have only ever known this previously terrified and traumatised animal as a calm, relaxed and affectionate ideal pet.

Toby

Toby was a six-and-half year old Cairn Terrier owned by Mavis and Don, a retired couple. They had got Toby from a dog home at the age of seven months to replace their previous Cairn that had died.

When Mavis called us, we could hardly hear her voice on the phone for the level of background noise. We guessed the problem immediately: a barker!

When Toby had been put out into the garden and we could talk, Mavis told us he had been barking constantly for six long years. It was driving them crazy. They had tried herbal tranquillisers, with no effect, and they had had his diet and general health checked with two different vets: all clear. Mavis told us (rather to our horror) that we were the last resort and if we couldn't help, they were going to have him put to sleep.

Our visit to the house was a lengthy one, as it was almost impossible to maintain a conversation. The minute anybody spoke, Toby started barking deafeningly in an infuriated, high-pitched tone that had everyone putting fingers in their ears. They said he was like this all the time. He barked at everything: when they were on the phone, when they watched television, when they had visitors. Their house stood on a hill, with a fine panoramic view across a valley that could be admired from a large bay window in the main room. But if they stood looking out of that window for more than about five seconds, Toby would immediately come rushing up and go berserk, barking and snapping until attention was paid to him instead of the view.

Evidently, what we had here was a fairly extreme case of an attention-seeking dog. The owners described to us how they had felt sorry for Toby when they adopted him, and had probably pandered to him all the more because he reminded them of their old dog, Barney. Everything in the house was geared to him. He ate his meals before they did, always leaving a little in the dish that he would come back for later on (no one touches the pack leader's food). He had his own chair, which he kept reserved with threatening growls but rarely sat on because he preferred the sofa, and slept on their bed at night. It was impossible to keep him out of any of the rooms, as it was an ultra-modern house with sliding doors everywhere which he had learned to open with his nose. In an attempt to create a space in their home where they could get some peace, Don had once tried blocking one of the doors with an armchair. Within hours, Toby had scratched all the paint off the door and ripped all the stuffing out of the chair in his outrage at being denied access. Basically, he ruled the roost. It was impossible to make him live in the garden in a kennel, as his incredibly noisy, rapid-fire barking had already provoked complaints from neighbours and a visit from the police.

Phew! We had a feeling that starting the case with too many training measures would probably have a negative effect. Firstly, Toby's behaviour would certainly get worse initially in protest, and the people might give up; and secondly we were afraid of confusing the elderly couple with a huge list of complex rules and regulations to follow. Instead, we offered a few simple suggestions such as eating all meals before he did, taking away all food that he left uneaten, and trying to ignore him as much as possible. We decided that the only way to break the case was with the flower remedies.

As we felt it covered the same ground as both **Vine** and **Chicory** (domineering, manipulative, self-seeking and sulky), we prepared a treatment bottle with the Australian remedy **Isopogon**. We added **Chamomile** for Toby's very vocal way of demanding attention, and **Dagger Hakea**, which we have long favoured for states of resentment. As a general aid against his anger and vexation, we put a few drops of Bach flower **Holly** in for good measure.

We left it at that for a few weeks, crossing our fingers whenever we thought of 'barking Toby'. After a month or so, Mavis phoned us. The first thing we noticed was that it was possible to speak to her without shouting. She reported that Toby was at that moment fast asleep on his chair.

Two weeks after starting on the drops, the dog had started to relax. He had initially been quite annoyed when they ate before him, but then it gradually started not to bother him and he would just lie and wait for his food in silence. He seemed less insistent on being the centre of attention, and when they ignored his demands as we had advised, he would whimper a little and then go and lie down. When they looked out of the window and stood admiring the view, he would still sit bolt upright and take notice, sometimes coming over to make himself seen – but the barking and the snapping had stopped.

Don and Mavis were amazed at the silence in the house. They realised that they had been shouting at each other all the time, and normal conversation over dinner felt like a whisper. They could watch TV in peace and quiet, only occasionally interrupted by a moan from Toby.

We told them that we were now at a stage in the case where we could really get to work on implementing changes that would completely eradicate Toby's bad behaviour. We had in mind banning him from sleeping on their bed, getting him a crate to sleep in, controlling his access to all rooms, etc. But they told us they were happy with the progress they had made, which far exceeded their expectations. We agreed to leave it at that, and they promised they would get in touch with us if ever the problems should return.

They never have.

Slug

Mary and Joe adopted Slug from an animal sanctuary, where he had been born a year before. He was a 'Heinz 57' mongrel, with the mixed genes of Staffies, Collies, and a bit of Jack Russell, in his lineage. Slug's problem, Mary and Joe explained to us somewhat enigmatically, was that he didn't have a problem. When we saw the dog we understood what they meant.

He had earned his name whilst still at the sanctuary, on account of his lethargic, sluggish ways. Nothing seemed to matter to him; he would lie around all day, not taking any interest in his surroundings. Not even food would excite him very much, and he just ate enough to keep himself alive. Both the sanctuary and his new owners had him checked by vets, just in case there might be something wrong with him which was making him act this way. But all tests proved negative. One vet said his only illness was 'lazyitis'.

Mary and Joe had read a little about flower remedies, and wondered if there was anything we could do to buck him up. It wasn't as though he were suffering, but he didn't seem to be getting much out of life. He would not respond to playfulness, and seemed old before his time.

For once, here was an ex-sanctuary dog with a well-documented history. Because they'd known him all his life, the owners there were able to tell us with certainty that his life had been completely uneventful. He had never experienced fear, never been in a fight, and never had any sort of trauma whatsoever. It couldn't be a case of separation anxiety, as he'd never formed any bonds with his companions or keepers. When they had taken him home for the first time, he had just wandered nonchalantly over to his bed and flopped down as though he had lived there all his life. His case almost seemed like a form of depression, which had come over him for no apparent reason – yet Slug didn't appear unhappy at all. Just... uninvolved.

We decided we would try the Australian remedy **Old Man Banksia**, for lack of energy and sluggishness. Another Bush remedy, one we had never used with an animal before, suggested itself when we remembered they'd said 'old before his time': **Little Flannel Flower**. It was unusual that an apparently healthy young dog would show no interest in play. When we visited him, he was spread out flat in his bed, surrounded by squeaky toys, chocolate bones, 'Roll-a-treat' balls that hadn't been touched. Maybe the inner puppy was lying dormant in there, repressed, waiting to be coaxed out of him. Time would tell.

Nothing happened for a couple of weeks. Then we started getting a series of e-mails from Mary:

> 'Two days ago Slug began to nose his toy. We saw him later that night actually rolling on the kitchen floor. This is a first!'

> 'Slug is chasing balls. We can hardly believe it.'

> 'He actually looks as though he enjoys his walks now.'

> 'I went for a bicycle ride down the lane, and Slug ran along behind me barking. I've never heard his voice before.'

> 'I don't know what this stuff is, but it works!'

After the first treatment bottle was finished, Mary and Joe stopped the remedies. The only problem they have with the dog now is explaining to people why his name is Slug!

Fred

This seemed, on the face of it, like a very simple and straight-forward case. About six weeks before the New Year 2000 celebrations, a lady named Carol called to say she needed something for her year-old Border Collie cross, Fred. Fred was terrified of fireworks. A bang, even in the distance, would send him under the table in fright, and any kind of celebrations involving firework displays reduced him to a nervous wreck for days afterwards. Carol did not like to think what kind of effect the Millennium celebrations were going to have on the dog, with the whole town reverberating like a war zone.

We went to see the dog at home. We ran through his case with Carol, taking down all the details as usual. As always with a case like this, we asked Carol how she herself felt about fireworks. Was she scared of them? Carol hesitated – seemingly because she was surprised at the question – and said no.

One of the details that did come up was that Fred didn't have his own bed to sleep in or spend time in. At night he slept on the rug in front of the fire. We suggested that part of his nervousness might be down to not having his own place within the home where he felt protected. Every dog should have its own special place, and we recommended to Carol that she should get him a dog bed, or a crate, that should remain in one spot and be Fred's own sanctuary. We also said that when he displayed the fear, she should ignore him and act casually, rather than try to console him or pet him – which would reward and reinforce the negative behaviour. It would also be a good idea to try a few canine body-language tricks, known as 'calming signals'*, such as yawning and blinking, to show Fred that the pack leader wasn't afraid.

As well as suggesting these simple measures, we gave Fred a treatment bottle of **Dog Rose**, **Rock Rose**, and – although Carol could recollect no specific trauma in Fred's life that might have given rise to the fear – **Star of Bethlehem**.

With three weeks to go before the celebrations began in earnest, Carol phoned to say that there had been no significant improvement in Fred's condition. She had got him a bed, to which he had become attached. She thought he seemed slightly more settled as a result of that. But the main problem with the fireworks was still there. Every so often, people in the neighbouring streets were letting off bangers, and Fred was running off to cringe in his new bed and scared for a long time afterwards.

We were somewhat surprised that this combination of remedies didn't seem to be working. Was Carol giving the right dose? Was Fred spitting out the treated snacks? We decided on

---

* Turid Rugaas, author of the short but excellent book *On Talking Terms with Dogs*, is an authority on the subject of canine calming signals. Details of his book are given on page 89.

an early second visit, to see if we could figure out this perplexing case.

It was a weekend evening when we saw Fred and Carol again. We checked that she had been giving the remedies according to instructions, and that he was taking them. No problem there. We asked a few more questions, but were, in truth, running out of ideas. It seemed as though the remedies, despite being well indicated for the case, just weren't having any effect.

At one point, Carol offered to make us all some tea, and while she was in the adjoining kitchen we were exchanging baffled glances as we petted Fred. Then, suddenly, there was a loud **BANG** from outside – someone letting off a firework a few gardens away. Fred dived for his bed, tail between legs, whimpering… and, at **exactly** the same moment, there was a crash from the kitchen that was almost as loud as the firework. Carol had dropped the tray with all the cups and teapot, plates of biscuits, sugar, all over the floor.

That was our answer. It turned out that Carol herself was terrified of fireworks after all. When we asked her why she hadn't told us, she burst into tears and told us how ashamed she was that she couldn't admit her fear, that she felt she should have got over it by now but was too weak. She had thought that if she could at least help Fred, it would be enough. But her fear was too overpowering, and she couldn't give Fred any sort of calming signals when she felt anything but calm herself.

So suddenly a whole new case had begun. Carol said that she was afraid of all loud bangs. When she was four years old, her elder brother had mischievously burst a balloon in her face as she slept, sending her into hysterics, and the fear had remained.

On this basis, Fred's new remedies were **Dog Rose** and **Rock Rose** as before, plus **Pink Yarrow** to help prevent his picking up and mirroring Carol's anxiety. Carol was given **Rock Rose** and the Australian remedy **Fringed Violet**. This one is indicated for especially deep shock that may be connected to violent frights and traumas in the distant past and has not been allowed to heal, instead lurking under the surface of the conscious mind to reappear at the slightest stimulus. Carol was very ashamed of her fear, and quite self-critical generally. We felt this merited **Pine**, for guilt, regret, and self-blame.

Result: we were out walking our dogs across the fields at

midnight on 31st December 1999, and we saw in the distance the bright flashes lighting up the night sky above the town, looking like news footage of an artillery battle. We wondered how our patients were coping in the middle of that.

A few days later, Carol called, with a lot to say. She had had some vivid and striking dreams shortly after beginning the remedies, where she recalled the balloon event in great detail. The first time she woke up trembling; but as it returned, it was less troubling. (We explained that this often happens with deep-acting trauma remedies, with negative unconscious content being confronted and released in dreams.) After the dreams, Carol went on, her tension at hearing a loud bang just seemed to melt away. She was still a little bit nervous and anticipating the next bang, but it was obvious that her state of mind was no longer negative enough to influence Fred's behaviour. He had lain quite relaxed in his bed all through the fireworks, occasionally looking up at her when there was a particularly loud one. He hadn't whimpered or acted scared a single time. His demeanour was generally more confident.

Buster

Buster was a ten-month-old Labrador owned by Debbie. When Debbie was sent abroad for a year by the company she worked for, she left Buster with her parents, Alan and Linda. Although they'd never owned a dog in their lives, they were quite happy to take him in, as they loved Buster and had always got on well with him during visits.

Buster seemed to settle in fine, and a few days went by. One morning he was having his breakfast in the kitchen when Alan walked past and accidentally nudged Buster's dish with his toe, moving it slightly. To his surprise and consternation, the normally placid dog let out a long, low growl and showed his teeth. Alan jumped back, and not knowing what to do, tried to pat Buster to reassure him that he wasn't trying to steal his food. Buster growled again, and this time Alan backed off completely and left the room.

He and Linda discussed it, and they concluded that Buster's

strange behaviour must be down to missing Debbie. Maybe they weren't giving him enough affection, they thought.

So they started letting Buster up on the sofa with them, and he happily lay across their knees, letting himself be cuddled for hours on end as they watched TV in the evenings. When they were eating their dinner, Buster would come and put his paw on their knee. They would look at each other and say: 'Aww, isn't that sweet?' and give him choice pieces of steak from their plates. Surely, thought Alan and Linda, all this VIP treatment would make Buster realise how much they loved him, and he need not be worried about his mistress' absence.

This went on for about a month, but it did not have the desired effect. Buster would now growl if they went anywhere near his food dish. He was also becoming possessive of the sofa, and one time he growled at them and wouldn't let them sit on it. Another time, Linda was sitting there reading, and Buster climbed up and parked his whole weight on her lap. With the book shoved in her face, she tried to shift herself into a comfortable position. Buster turned his head round and curled his lip at her, an inch from her nose! Linda was so terrified, she sat there rigid for an hour with terrible cramps in her legs, until Buster decided to move away.

At this point, they were aware they had something of a problem. They got in touch with a vet, who referred them to a dog behaviourist. This lady visited the house, asked them lots of questions, and gave them lots of very good advice on how to re-establish control over Buster. She explained that they had created a 'pack leader paradise' for the dog, starting with that first time when Alan backed away and then tried to pat the dog. Without realising it, Alan had been rewarding Buster for his bad behaviour, and it had escalated from there. What was needed was a programme of 'dog demotion', a simple list of measures they could implement to show Buster who was in charge of the house. No more going up on the sofa, no more being fed at the table, no more ridiculous pampering! She also suggested that he be taken to weekly dog training classes.

In theory, this should have sorted the problem. But Alan and Linda had now become so afraid of Buster that implementing the changes in their routine with him was impossible. They just couldn't assert themselves over him, convinced he would turn round and bite. When they refused him food at the table, he barked loudly and stared them in the eye until they relented. And then he would demand more! They tried prodding him off the sofa with a broom handle (it would be funny if it weren't true!), but he attacked it.

What to do? By chance, a friend of a friend had seen us in connection with a vicious cat that had been cured with flower remedies. So, back to the vet for another referral, this time to us, and we went to see them.

It was obvious that before Buster's training programme could be started, something had to be done to get over the immediate barriers. Buster was given the flower remedies **Vine**, for his dominant grip on the household; **Snapdragon**, for his obviously growing urge to bite, and **Dagger Hakea**, in anticipation of the resentment he would probably feel when they started demoting him within their 'pack'. To help the readjustment and create a better harmony within the pack, we added **Quaking Grass** to Buster's treatment bottle.

Alan and Linda clearly needed something, too, to get them over their fear of the dog. They were both given the aptly named **Dog Rose**, to address their fear and lack of confidence, and **Centaury** to allow them to assert themselves more. They also said they felt guilty, first for having allowed their daughter's dog to get so out of hand, and second, for having to 'do this to poor old Buster'. (We often find that even in cases of flagrant bad behaviour, the owners feel guilty about implementing what they perceive to be overly disciplinarian measures against their dogs: *'He won't love us any more!'* This is a false belief.) And so they were given an extra remedy, **Pine**, to alleviate the sense of guilt.

Result: within two weeks, Buster had mellowed sufficiently for Alan and Linda to start implementing his training. He had stopped growling at them over his food, and although he was still pestering them at the table, his bullying and barking stopped sooner when they refused him. They were more able to assert themselves in small but important ways, such as walking through doorways ahead of him and various other tips in the 'dog demotion' programme the behaviourist had given them. Little by little, the dominant behaviour of this dog was reduced and controlled, until they had a normal dog again. As this is written, Alan and Linda are awaiting their daughter's return, confident that they can give her back Buster the same way they got him! Neither flower essences nor training could have achieved this fast and satisfying result on their own.

Suzy

Suzy is a Great Dane bitch, but when we first saw her she looked more like a Greyhound, and a really skinny one at that! She was taken in by Great Dane Care in Carmarthenshire after her previous owners, having obtained her because they liked the 'idea' of the breed, had subsequently decided she was too expensive to feed, and had effectively been starving her. She had been kept tied to a post at the bottom of the garden, very poorly cared for and was extremely stressed and nervous. When being moved to the sanctuary, she became aggressive and tried to bite her handlers; then later she bit someone who entered her kennel to feed her. Luckily, the bite was not too serious – only a warning.

On seeing the dog, it was very obvious that her aggression was based on fear rather than on any kind of dominant or controlling urge. The carers at the sanctuary observed that she seemed particularly afraid of a red leash, while a blue one did not seem to bother her so much. They wondered if she might have been beaten with a red lead or strap at some point. In any case, she was very severely traumatised. When we approached her kennel, taking care not to look her in the eye for fear of intimidating her, we noticed how tense she looked as she cowered at the far end of the enclosure, snarling at us with her tail tightly coiled up between her legs. The people at the sanctuary had told us the tail had been permanently planted in this position since she had arrived a few days before.

A couple of months before seeing Suzy, we had had the good fortune to attend a seminar held by Steve Johnson, the founder of the Alaskan Flower Essence Project. This seminar, held at the Living Tree in Hampshire, was important in that it doubled as the UK launch of a brand new flower remedy formula, called **Animal Rescue**. Steve had been working on this formula for the past decade, in conjunction with his colleague Andrea Freixida of BioFauna in Sao Paulo, Brazil. Andrea is a flower essence practitioner and biologist who works with wild animals at a safari park. The seminar featured a striking case of some abandoned circus lions that had been successfully treated using this formula, using it in spray form as the lions were initially too dangerous to approach. This case has been well documented and was even featured on Brazilian TV news! Andrea's team had made their own video, which showed graphically how the lions' behaviour had shifted from aggression, roaring and lashing out with their claws, to rolling on their backs and kicking their feet in the air with pleasure. An amazing sight!

We had a spray of **Animal Rescue** with us when we visited Suzy, as we had anticipated that it might not be safe to enter the kennel. We remained very quiet, and stood near the entrance to her kennel to show her we were not a threat to her. She approached every so often, but in a very fearful way, and would back off, snarling and growling at us with her tail still wrapped up tightly against her abdomen. After a few approaches, we gently sprayed a few puffs inbetween the bars of the kennel. She sniffed at it, and every so often we would mist some more of the formula in the air around her.

After less than five minutes, we were astonished to witness Suzy's tail no longer curled up between her legs but now suddenly straightening out and hanging down. She also stopped growling and snarling, and a look of softness, as though shedding muscular tension, was coming over her. It was the same reaction we had seen in Steve's video of the lions! Next time she approached us, we were able to feed her a biscuit with some of the formula sprayed on it. Then, after a few more minutes, Suzy felt comfortable enough to lie in her bed for a while, until we left.

We gave the rest of the spray to the sanctuary owners, and left it at that. A few days later, we began to get some feedback about Suzy's progress. The sanctuary doubles as a training centre for students of animal care, and we learned that some students had been able to enter Suzy's kennel freely and pet and handle her, without any sign of fear or aggression on her part. It was also possible to feed her meals to her with no concerns about safety.

A week later, we were returning to the kennel when we met

two young students walking Great Danes along the country lane. One of the dogs bore a resemblance to Suzy, but as we drove past we decided it couldn't be her, as this dog was far too plump and healthy-looking! On arriving at the kennel, however, we discovered to our pleasant surprise that the dog we had seen **was** Suzy. She was now completely placid with handlers and proving to be a loving, affectionate and sweet-tempered dog.

We wish there were more to report on this case, but the simple fact of the matter is that Suzy never looked back after that first treatment with the **Animal Rescue** formula. She required no further treatment, and has now been found a suitable home with responsible owners.

To us, this case highlighted two major points: one, that a fundamentally good dog with a real capacity for affection and loyalty can easily be transformed into a potentially dangerous, and suffering, animal by human negligence and cruelty (either intentional or unintentional). And second, it showed once again the amazing ability of flower remedies to heal even the deepest emotional cuts that such cruel and irresponsible treatment can inflict on these wonderful animals.

6 Using the Remedies

Choosing remedies

We hope that this little book will have given you pretty clear ideas about which remedies to choose for doggy problems. All that you need to do is think carefully about whatever problem it is that you have encountered, and match it to the best-indicated remedy or remedies. Using the different sections of this book, you can cross-refer and check that your chosen remedies are close-fitting to the problem. **You can select up to six remedies –** some combinations use more than six – but we often try to whittle our selections down to four or five, as you can see from the case studies. 'Whittling' involves trying to focus on the key issues of the case and cutting out semi-indicated remedies that don't really fully apply. You can always change your selection later, if necessary.

If you've never used the flower remedies before and it seems daunting – relax. Even if you choose entirely the wrong ones and give those to your dog, you cannot possibly cause any negative effect. As we've said elsewhere in this book, a wrong remedy just 'bounces off'; the system will only respond to a remedy that it needs. Using flower remedies really is simple, and yet produces such positive results – definitely a skill worth acquiring. With a little practice, you will soon be able to pick out selections of remedies very quickly and efficiently.

'Stock' bottles, 'treatment' bottles and dosages

Flower remedies can be taken either from a stock bottle or a treatment bottle. The little bottle you buy from your supplier is known as the stock bottle. The flower remedy it contains is a dilution of the original mother essence in brandy. Though less dilute than the remedy in a treatment bottle (*see below*), it's not concentrated as such, and it is perfectly fine to take drops straight from there. As for dosage, we are talking about a completely different concept from allopathic, or conventional, medicine – there are no real rules. It depends on the application. In acute or emergency cases you could take (or give, in the case of an animal) several drops at a time, every few minutes until improvement was seen, and then lessen that to every few hours for a day or so until the trauma, fright or whatever it was had been soothed away. In regular daily treatment, **use four drops, four times a day** as a rough rule of thumb. Consistent and regular intake are more important than taking a certain number of drops per day.

Treatment bottles, sometimes also known as dosage bottles, are something you can easily learn to make up yourself. Their advantages are basically twofold: firstly, they allow you to carry a combination of remedies all in one bottle so you don't have to fiddle about with a whole pocketful; secondly, they will save you money as much less of each stock remedy gets used up.

To make a treatment bottle, get hold of a 30ml (or thereabouts) amber glass dropper-bottle, which will be a larger version of the one you've got your stock remedy in. (We get our bottles from Ainsworths and have included their address at the end of the book.) Once you've got the bottle, you'll need some flat, not sparkling, mineral water and a little bit of brandy to act as a preservative for the water. Alternatives to brandy are cider vinegar and vegetable glycerine. But beware of using vinegar-preserved remedies for animals. They can smell it a mile away and get very suspicious. We favour the traditional brandy method. The alcohol content is very insignificant, and cannot cause any problems with your pet.

Fill the bottle about quarter full with brandy and then top up about four-fifths with water. You should have enough space left to add four to six drops of each remedy you have chosen. Once they are in, screw the cap on tight, and give the bottle a shake to mix up the contents. Many people believe in shaking or tapping the bottle every time before use, as a means to energise or potentise the remedies inside. Before you laugh, bear in mind that this 'succussion' is a vital part of the preparation of homoeopathic medicines and has been employed in homeopathic pharmacies for two hundred years.

If you use four drops, four times a day from the treatment bottle, it will last three to four weeks. Taking flower remedies in this fashion, it's possible to make your original stock bottles last a very long time and make flower remedies a highly economical business.

How do you know when to stop with the remedies? As it's impossible to overdose, you can feel free to carry on for as long as required. It should not be necessary, though, to go on with the treatment for more than a few weeks.

A structured way of doing it is to use up a treatment bottle (3–4 week cycle), pause for a week to check and compare results and see if further treatment is necessary, then move on with another treatment bottle, and so on. Acute problems and passing bad behaviour will require the shortest length of treatment if backed up wisely with practical measures. Traumatic memories may take a little longer, although in many cases are healed with spectacular speed. Ingrained personality traits and negativity in the dog's character may be the last thing to vanish entirely, and may require ongoing treatment for a few months. These are the most likely to re-occur in future, but in our experience they will respond better and faster to flower remedies with repeated cycles of relapse/treatment. Eventually, they often disappear altogether. The remedies are curative, not just palliative like most medicine.

Administering remedies to dogs

The usual method people use for taking remedies themselves is straight from the bottle (stock or treatment) onto the tongue. With animals, however, although it is possible to give remedies by the same method there is the concern that an animal may bite off and swallow the glass dropper tube, which could be fatal to it. In our opinion it is always better with animals to use either water or food as a carrier for the remedy.

You will be pleased to know that dogs are generally easier to give flower remedies to than other animals. Some horses can get very suspicious of treats that you may have doctored, and spit them out in your face. Cats may suddenly decide they don't like smoked salmon after all when it has a couple of drops of **Holly** on it. Dogs are the best.

Our method with our own dogs, and what we recommend, is just to put some drops onto a small piece of bread, call the dog over and make a big fuss as though he were getting something really special, and then pop it in his mouth. They nearly always gulp it down greedily. This is a very easy thing to do four times a day. Alternatively, drop the drops into drinking water – but if you have more than one dog you may find your patient getting less of the remedy than he should.

Another way is to put the drops in with a dog's meals, but as most dogs eat only one or two meals a day, you will need to have another way of giving the remaining two daily doses. If you're really stuck, instead of giving four drops four times, give eight drops twice for extra convenience.

Storage and shelf-life of remedies

Your remedies are quite resilient and will not really need much looking after. Do keep them away from electrical equipment, stereos, computers and the like. We wouldn't carry one in the same pocket or bag as a mobile phone, just in case the radiation energy might corrupt the subtle imprint of the flowers.

At home, store the remedies in a cool, dark place. If you take these simple precautions, flower remedies should last a long time. Remember, energy lasts much longer than matter.

Suppliers of flower remedies

Bach Flower Remedies

A. Nelson & Co. Ltd are the company that took over production of the official, trademarked Bach Flower Remedies product and name from the Dr Edward Bach Centre in 1991. Their brand remedies, bearing the distinctive yellow label and Bach's signature, are very widely available from mainstream chemists in the UK, for instance the leading store Boots.

Nelsons may be contacted for information on where the original Bach flower remedies can be bought world-wide.

A. Nelson & Co. Ltd
Broadheath House, 83 Parkside, London SW19 5LP
Tel: 0208 780 4200

Ainsworths Ltd also produce a very similar product, although they are not legally entitled to use the Bach flower remedies' trade name.

Ainsworths Ltd
40-44 High Street, Caterham, Surrey CR3 5UB
(Orders/accounts/books)
Tel: 01883 340332

(Ainsworths will also sell you empty dropper bottles for the purpose of making up treatment bottles.)

There are also some smaller producers who make up lesser quantities of flower essences by hand, exactly in accordance with Dr Bach's methods.

Crystal Herbs Ltd
1D Gilray Road, Diss, Norfolk IP22 4EU
Tel: 01379 642374
Email: Orders@crystalherbs.com
Website: www.crystalherbs.com

The Healing Herbs of Dr Edward Bach
Healing Herbs Ltd, PO Box 65, Hereford HR2 0UW
Email: healing-herbs@healing-herbs.co.uk
Website: www.healing-herbs.co.uk

Either of these smaller companies will also supply overseas.

Australian Bush Flower Essences/FES Californian Flower Essences/Alaskan Flower Essence Project Essences
These, as well as many other families of flower remedy, can all be ordered in the UK from:

IFER (The International Flower Essence Repertoire)
The Living Tree, Milland, Liphook, Hants GU30 7JS
Tel: 01428 741 572
Email: flower@atlas.co.uk

IFER will post overseas, but readers in the USA can obtain flower remedies from:

Flower Essence Pharmacy
6265 Barlow Street, West Linn, OR 97068 USA
Email: info@floweressences.com

7 Dog Demotion Programme

If you have read this book you will have seen that we often recommend this sort of programme as a means of complementing flower remedy use. It is laid out here for your convenience.

The ideas within the dog demotion programme are very commonly used in dog training. It is designed to do exactly what it says – to reduce the dog's rank within the hierarchy of the human pack. Dominance is the cause of a large proportion of dog behaviour problems, and following these rules will help reverse and eradicate many of the bad behaviours that have appeared as a result of a dog getting above himself in your household.

THE PROGRAMME

The following information is designed to help you, the dog owner, to lay down basic ground rules that will go towards ensuring that your dog remains within your control at all times, respects your authority and is happy and well-balanced.

First, a note on consistency.

Consistency is one of the most important things to bear in mind. If you have decided you want to modify the behaviour of your dog, it **must** be an 'all or nothing' effort. Everybody in the household must be committed to the rules of the programme. There is no point in half-measures, as these will prove counterproductive. It is extremely important that everybody be consistent with permission and denials. For instance: if you don't want the dog to occupy furniture intended for humans, or play with household objects such as shoes, but someone else in the family or household allows this behaviour, two things may happen:

1. It will undermine your authority, and you will lose all the progress you have made so far. Thus, any future attempt to assert your authority may result in nasty confrontation, as the dog has learned that it has ways of overcoming human authority.

2. It will confuse the dog. A confused dog is not only unhappy, but can also become aggressive. The law in most countries is very clear on aggressive behaviour, with many dogs being destroyed and many owners being heavily prosecuted each year.

Do not make the mistake of thinking that if others give into the dog and allow him to have his way, he is going to love them more and you less. A dog is happier when he knows his rightful place in the pack, and when he can easily recognise who the pack leader is. In the case of the domestic dog, the pack leader is EVERY human in the household. This message must be clear and unequivocal.

GROUND RULES FOR DOG DEMOTION

- When any member of the household comes home, they must walk past the dog, avoiding eye contact or touch, and greet other humans first. **Then** they can say a **quick** hello to the dog. If the dog has gone off to sulk, ignore it.

- **Do not** let the dog precede you through any doorway or narrow space. This is a privilege reserved for the pack leader. (Remember that these concepts are uppermost in the dog's mind, especially if the dog has a dominant urge.)

- **Do not** approach the dog to give cuddles, especially when he is lying down.

- **Do not** respond to the dog when he tries to get attention from you, e.g. coming up and lifting your arm with his nose. If he does try, keep sending him away. Every so often, unexpectedly call the dog to you, make a fuss of him, and send him away again.

- If the dog lies in a doorway, at the top of the stairs or anywhere where he is blocking people's path, **do not** step over him. Tell him to move; if he doesn't do this quickly, give him a gentle shove with your foot if necessary. In the dog pack, no lower-ranking member would ever dare make the leader move. When the dog sees you move round him or step gingerly over him, he feels he has been elevated in rank. (It is not that dogs are scheming creatures. They only exercise hierarchy for the survival of their pack, the way an army officer must keep discipline for the good of his unit.)

- Teach the dog to leave a room on command, even if it is only for a few minutes before letting him back in.

- Prepare the dog's food in front of him, then leave his feeding until **after** you have eaten. (If feeding in the middle of the day, a coffee and biscuits counts as a symbolic 'leader's share' of the food.) When you do offer the dog his food, make it work for it. Give a command such as 'sit'. The dog does not get his food until he sits.

- If he doesn't eat all his food within five minutes, take it away and do not offer any more until the next allocated mealtime.

- Keep control of all toys, balls, etc. **Do not** allow them to lie scattered around the home or garden. You initiate all play when the dog is not expecting it. After the game, which you bring to an end, take possession of all playthings and put them away out of the dog's reach or sight until next time.

- Give the dog a daily grooming. Keep sessions short to begin with, and if you like reward the dog with snacks. Gradually build up the time spent grooming, and get others to share in the exercise. This gets the dog used to anyone handling and brushing it.

• Last but not least, **do not let the dog sleep on your bed**. If you allow this, in the dog's eyes you have elevated him to a position at least equal to yours in rank and possibly higher. If you find him on the bed, order him off immediately in a very loud and authoritative voice. **Give no reward** for obeying you (not even verbal praise), but stop your pretended 'anger' immediately as soon as the dog is off the bed. Then he will see how to switch off/avoid your annoyance by behaving well.

If everybody follows these basic rules, many of your dog's behavioural problems will resolve themselves. Remember that how the dog has been allowed to interpret his place in the human pack is probably what has initiated the problem in the first place. This groundwork paves the way for further training and complements natural behaviour modification/treatment such as flower remedies.

8 Where Can I Learn More?

A few other books have been written about using flower remedies for animals. Until now these have focused on the Bach remedies, but as research and development progress hand in hand with growing popularity, there will be more books in the future that take the wider approach we have started to take in this book, and widen it further and further. However, the Bach remedies on their own are an excellent start for anyone wishing to learn about using flower remedies for animals. Here is one reading suggestion:

Bach Flower Remedies for Horses and Riders
Martin J. Scott with Gael Mariani
Kenilworth Press, Buckingham, 1999

And it would be a good idea to get hold of at least one or two books purely on dog behaviour, psychology and training. We would recommend:

Think Dog!
John Fisher
Blandford, UK, 1990

The Perfect Puppy
Gwen Bailey
Hamlyn, UK, 1995

Training Your Best Friend
John Rogerson (with a section on Contact Learning by Julie Sellors)
Stanley Paul Ltd (Random House), UK, 1993

On Talking Terms with Dogs: Calming Signals
Turid Rugaas
Legacy by Mail Inc., USA, 1997

The Trouble-free Dog
Robert Alleyne
Robert Hale Ltd, UK, 1999

As well as books, there are organisations one can join that provide a forum for all those interested in this subject.

The Society for Animal Flower Essence Research (SAFER)
Tel: 01267 281761
Email: flower_essences@uku.co.uk

SAFER is based in Wales, UK, but has members in England, Scotland, Republic of Ireland, France, Canada, and many States of the USA. Its aim is to promote the use of, and interest in, flower remedies for pets by their owners and by professionals. Yearly membership is inexpensive. Members receive a quarterly newsletter and are invited to contribute ideas and articles. Several major producers of flower remedies, including Patricia Kaminski of the Californian flower essences and Steve Johnson, the creator of the Alaskan collection, are involved in contributing articles and case studies for the Society.

The Living Tree (IFER) in Hampshire produce a newsletter that goes out to all customers on their mailing list and is full of information on new flower essence products, books and events. Additionally, they hold many seminars and talks on their premises with some of the world's leading exponents of flower remedies. Many of these speakers, such as Steve Johnson of the Alaskan Flower Essence Project, are expert in animal treatment. The address and telephone number of this company are listed in Chapter 6.

Index

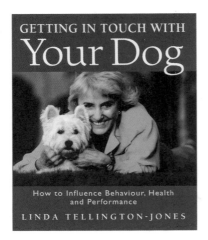
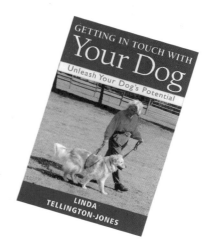